SpringerBriefs in Applied Sciences and Technology

Nonlinear Circuits

Series editors

Luigi Fortuna, Catania, Italy
Guanrong Chen, Heidelberg, Germany

W0230281

SpringerBriefs in Nonlinear Circuits promotes and expedites the dissemination of substantive new research results, state-of-the-art subject reviews and tutorial overviews in nonlinear circuits theory, design, and implementation with particular emphasis on innovative applications and devices. The subject focus is on nonlinear technology and nonlinear electronics engineering. These concise summaries of 50–125 pages will include cutting-edge research, analytical methods, advanced modelling techniques and practical applications. Coverage will extend to all theoretical and applied aspects of the field, including traditional nonlinear electronic circuit dynamics from modelling and design to their implementation.

Topics include but are not limited to:

- nonlinear electronic circuits dynamics;
- Oscillators;
- cellular nonlinear networks;
- arrays of nonlinear circuits;
- chaotic circuits;
- system bifurcation;
- chaos control;
- active use of chaos;
- nonlinear electronic devices;
- memristors;
- circuit for nonlinear signal processing;
- wave generation and shaping;
- nonlinear actuators;
- nonlinear sensors;
- power electronic circuits;
- nonlinear circuits in motion control;
- nonlinear active vibrations;
- educational experiences in nonlinear circuits;
- nonlinear materials for nonlinear circuits; and
- nonlinear electronic instrumentation.

Contributions to the series can be made by submitting a proposal to the responsible Springer contact, Oliver Jackson (oliver.jackson@springer.com) or one of the Academic Series Editors, Professor Luigi Fortuna (luigi.fortuna@dieei.unict.it) and Professor Guanrong Chen (eegchen@cityu.edu.hk).

More information about this series at http://www.springer.com/series/15574

Ferdinando Taglialatela Scafati
Mario Lavorgna · Ezio Mancaruso
Bianca Maria Vaglieco

Nonlinear Systems and Circuits in Internal Combustion Engines

Modeling and Control

 Springer

Ferdinando Taglialatela Scafati
Automotive and Discrete Group
STMicroelectronics
Arzano
Italy

Mario Lavorgna
Automotive and Discrete Group
STMicroelectronics
Arzano
Italy

Ezio Mancaruso
Istituto Motori
National Research Council (CNR)
Napoli
Italy

Bianca Maria Vaglieco
Istituto Motori
National Research Council (CNR)
Napoli
Italy

ISSN 2191-530X   ISSN 2191-5318   (electronic)
SpringerBriefs in Applied Sciences and Technology
ISSN 2520-1433   ISSN 2520-1441   (electronic)
SpringerBriefs in Nonlinear Circuits
ISBN 978-3-319-67139-0   ISBN 978-3-319-67140-6   (eBook)
https://doi.org/10.1007/978-3-319-67140-6
SpringerBrief: Copyright retained by authors.

Library of Congress Control Number: 2017952523

Printed on acid-free paper

This Springer imprint is published by Springer Nature
The registered company is Springer International Publishing AG
The registered company address is: Gewerbestrasse 11, 6330 Cham, Switzerland

Preface

Today, cars include approximately one-third of their value in electric and electronic components, and the mobility paradigm is being transformed toward "the smart driving" concept, with the aim of enhancing the driver experience, improving the safety, supporting the connectivity and automated driving, but also lowering environmental impact.

The engine is only a part of this complex structure of the vehicle.

However, internal combustion engine remains the main source of energy; its function is not different from that of the first prototypes, which is to convert the chemical energy contained in the fuel in mechanical power. This process involves many complex thermo-fluid dynamic phenomena affected by nonlinear dynamics: intake air motion, air–fuel mixture dosage, combustion process itself, knock and misfire occurrence, particulate particle formation, just to cite few of them.

The challenge during all these decades has been to optimize the combustion process in terms of engine efficiency and pollutant emissions reduction, also to comply with the more and more strict governmental rules.

The challenge is still open.

With this work, the authors want to refocus the attention of academic and industrial automotive experts on nonlinear processes in internal combustion engine, analyzing specific nonlinear conditions, providing original modeling description and effective control solutions able to compensate these nonlinear dynamics.

Chapter 1 is aimed at describing the use of Artificial Neural Networks and Expert Systems in engine applications. Artificial intelligence techniques allow to solve highly nonlinear problems offering an alternative and effective way to deal with complex dynamic systems. Air–fuel ratio (AFR) modeling and control is a typical highly nonlinear problem where a huge number of interconnected parameters needs to be considered and controlled (amount of fuel injected, residual gas fraction, wall wetting are some of the parameters that have to be processed). In this context, we propose a neural network and fuzzy logic approach for AFR modeling and control.

In Chap. 2, advanced non-interfering diagnostics based on optical spectroscopy are presented. Optical diagnostics allow to take a look in what really happens in the cylinder in terms of flame propagation, gas turbulences, and pollutant formation. In other words, most of the phenomena occurring during the combustion process. The evaluation of these nonlinear phenomena is the key point to design effective control solutions able to optimize engine combustion in terms of engine power, efficiency, and emissions.

Nowadays, great attention is paid to the impact of particulate matter (PM) emitted from vehicles on the environment and, in turn, to the negative effects that it has on human health. Pollutant particles are classified according their diameters in micron (PM10, PM2.5, etc.); smaller the particles are, more dangerous for human health they are as they penetrate more easily the cell membranes. The chemical nature of the emitted particles as well as the number and size depends on engine type and its operating conditions. In Chap. 3, the authors deal with the particulate emission reduction problem, suggesting a real-time approach to model the number and size of emitted particles.

The parameter widely considered as the most important for diagnosis of the combustion process in internal combustion engines is the cylinder pressure. This signal represents, in fact, the most direct signal available for engine control. However, in-cylinder pressure direct measure involves an intrusive approach to the cylinder using expensive sensors and a special mounting process. For this reason, several alternative methods for combustion diagnosis have been suggested in literature. In Chap. 4, we propose a method for advanced and non-intrusive combustion diagnosis using the vibration signal produced by the combustion process on the engine block. Real-time engine control architectures that use this signal are also investigated.

Recently, more robust and cost-effective in-cylinder sensors have been developed, and their usage in mass-produced vehicles now appears more feasible. These new types of pressure transducers are generally integrated in the glow plugs, in the spark plugs, or into the injector valves. Chapter 5 provides an overview of the main applications of cylinder pressure signal in engine modeling and control.

In Chap. 6, the nonlinear phenomena correlated with the injection process in GDI engines are analyzed. A complete description of the injector nonlinear dynamics is provided, and an effective compensation is proposed.

Latest emission regulations, in fact, strongly push toward a reduction of fuel consumption in order to reduce CO_2 emissions because of their effect on global warning. Gasoline direct injection engines, together with fully electric and hybrid vehicles, are the best candidate to satisfy the imposed limits. GDI engines, in fact, can work in stratified operations allowing stable combustions with ultra-lean mixtures that allow a strong reduction of toxic emission coupled with fuel consumption reduction. GDI stratified operation needs the use of multiple fuel injections, splitting the quantity of injected fuel into several and shorter shots in order to reduce the cylinder wall impingement.

However, small injections force solenoid injectors to work in *ballistic* mode, i.e., the injection pulse width is cutoff before the valve fully lifts up, causing a highly nonlinear correlation between electrical command pulse width and the actual amount of injected fuel. We present a close-loop control able to manage and compensate the ballistic behavior.

We wish the material collected in this Brief can stimulate the interest of young undergraduate and graduate students, researchers both academic and of industry pushing them to develop research projects exploiting nonlinear dynamic problems in combustion engines.

To conclude, we would like to thank all the colleagues that gave their fundamental contribution in the achievement of the results presented in the Brief.

Ferdinando Taglialatela Scafati
Mario Lavorgna
Ezio Mancaruso
Bianca Maria Vaglieco

Contents

About the Authors

Dr. Eng. Ferdinando Taglialatela Scafati got a master degree in Chemical Engineering at the University of Naples "Federico II" in 2001. In 2002, he joined STMicroelectronics working as a researcher in the Soft Computing Group— Corporate R&D. In this period, he focused his activities on the design of models for automotive systems and on the use of soft computing techniques for IC engine control. In 2005, he moved to the Automotive Product Group of STMicroelectronics where he currently works. Here, he is involved in the design, implementation, and validation of engine control systems for traditional and innovative power trains. He is a Member of Technical Staff of STMicroelectronics as recognized expert in the field of electronic control of powertrain systems and engine management. He is author of several patents and publications in peer review journals, and he collaborates with automotive companies, universities, and international research centers.

Mario Lavorgna was born in June 26th, 1957 in Rome, Italy.

He got a master degree in Physics in 1982 at University of Naples, Federico II. He worked until 1987 as a researcher in General Relativity and Cosmology at University of Naples, Federico II, publishing several scientific papers on international scientific journals.

He has worked since 1987 at STMicroelectronics, where he drives R&D activities in advanced microelectronic applications, including Fuzzy Logic, Neural Networks, Soft Computing methodologies in different applicative areas: consumer, industrial, robotics, automotive.

Actually, he is in charge of Mass Market System and Application for Automotive Group at STMicroelectronics, in Naples, Italy, developing electronic control solutions for the next generation of powertrain, cooperating with key automotive companies and international research centers.

He is author of several patents, technical books, and scientific papers.

Dr. Ezio Mancaruso is Ph.D. graduated in Mechanical Systems Engineer, is a researcher of the National Research Council (CNR) and made his activity in the Istituto Motori (IM). Principal areas of activity were fluid dynamics, thermodynamics and combustion in diesel engines and fuel–engine interaction. Ezio focused his research activity on the application of the optical diagnostics and advanced combustion sensor to study in-cylinder phenomena and to develop new control system for ICE.

In particular, he carried out an intensive experimental research activity on the fluidynamic, injection, and combustion fields of the ICE using optical diagnostics as well as on the application of advanced sensors like accelerometer and ionization c current system to diesel engines. He has a good experience as reviewer for several ISI journals. He is co-author of more than 100 publications on peer review journals and book chapter. He is scientific responsible for workpages in EU, and coordinator and scientific responsible of several projects.

Bianca Maria Vaglieco is Director of Research of Istituto Motori-CNR, she oversees several engine laboratories. She is involved in several activities related to the experimental and theoretical study of the thermo-fluidynamic process in reciprocating internal combustion engines. The main activity concerns in the experimental investigation of fundamental physical and chemical processes occurring in CI and SI engines by means of non-intrusive diagnostics such as innovative sensors and optical systems. She has developed advanced non-intrusive optical diagnostics for analyzing chemical and physical phenomena in the cylinder and exhaust in order to contribute to the improvements of engine performance and emission. She is author of more than 340 papers including peer reviewers journal papers, book chapters, and international conference and patents.

Chapter 1
Artificial Intelligence for Modeling and Control of Nonlinear Phenomena in Internal Combustion Engines

Artificial intelligence techniques allow to solve highly nonlinear problems offering an alternative way to deal with complex and dynamic systems with good flexibility and generalization capability. They are widely used in several areas ranging from power system modeling and control to medicine and social sciences. Because of their good ability to model nonlinear phenomena together with their relatively simple application procedure, artificial intelligence systems have found an increasing usage in the modeling, diagnosis, and control of internal combustion engines. The most used techniques include Artificial Neural Networks (ANNs), Genetic Algorithms, Expert Systems, fuzzy logic and hybrid systems, with several combinations of two or more of these.

The present chapter aims to describe the use of artificial intelligence in some engine applications where the inherent nonlinear nature of the process dynamics requires alternative approaches to guarantee a more accurate control action. A special focus will be kept on the use of Artificial Neural Networks and fuzzy logic techniques.

1.1 Neural Networks Architectures for Engine Applications

Neural networks provide a wide range of functions that can be used in the field of engine control. They can be used for example to train *black box* process models of various engine subsystems with few a priori theoretical knowledge. In this way, difficulties that appear when applying classical techniques on complex nonlinear systems are suppressed.

A basic characteristic of neural networks is that of emulating the structure of human brain and, in particular, its ability to learn from experience without actually modeling the physical and chemical laws that govern the system [1–3]. In general terms, an Artificial Neural Network is a computational system able to store and

© The Author(s) 2018
F. Taglialatela Scafati et al., *Nonlinear Systems and Circuits in Internal Combustion Engines*, SpringerBriefs in Nonlinear Circuits, https://doi.org/10.1007/978-3-319-67140-6_1

Fig. 1.1 Example of a
neuron model

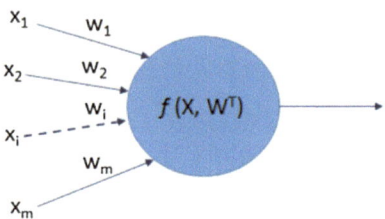

utilize knowledge acquired through experimenting [4] and it can be defined as an interconnected assembly of simple computational elements called *neurons* or *nodes*. Each element receives the inputs from neighboring nodes, sums all these contributions, and produces an output that is a nonlinear function of the sum. The signals flowing on every connection are appropriately scaled by programmable parameters called *weights*. In Fig. 1.1, an example of neuron model is depicted. In Fig. 1.1, x_i represents the ith input (i.e. the ith component of the vector X), w_i is the relative weight at the ith input (ith component of the vector W), and $f(W^T, X)$ is a function, known as *activation function* whose value, calculated as weighted sum of the inputs, represents the neuron outputs.

An important part of the modeling with neural networks is the so-called training of the network (*learning procedure*). This latter is the process that, using different learning algorithms chosen according to the network structure and type, assigns values to the network parameters (in particular, to the connection weights) in order to minimize the error between the outputs of the neural model and the correct or desired outputs.

The most popular learning algorithms are the *back-propagation* and its variants, which are generally applied to multi-layer feedforward networks having differentiable activation functions. The back-propagation algorithm is an iterative algorithm that updates the values of the network interconnections such that a total square error is optimized on a set of input/output data. The error can be expressed by:

$$E = \frac{1}{2} \sum_p \sum_i (t_{i,p} - y_{i,p})^2 = \frac{1}{2} \sum_p E_p, \tag{1.1}$$

where p is the number of data supplied during the learning, i is the number of outputs (i.e. of neurons in the last layer), t is the desired output value, and y is the corresponding value calculated by the network. Initially, the training is performed by assigning random values to the weights $w_{i,j}$. With every iteration, one of the training set samples is provided to the neural network and the error committed by each neuron output is calculated. Then, the gradient algorithm for back-propagation of the output error is applied backward through the network updating the value of the weights, according to the formula:

$$\Delta W_{ij}(t) = -\varepsilon \frac{\partial E_p}{\partial W_{ij}}, \tag{1.2}$$

where ε is a parameter chosen by the user and called speed of the learning.

The training of all the patterns of a training data set is called an *epoch*. The learning procedure uses a number of epochs that allows to obtain a sufficiently low error or an error that no longer decreases. This latter case indicates the incapacity of the network to solve the problem.

The learning procedure must be carefully designed. The training data should cover all nonlinearities and should contain information spread evenly over the entire range of the system. This allows to avoid significant model failure if the neural network model is used in a region where an insufficient amount of data is supplied. Moreover, for a good predictive ability of an ANN, it is important that the training and the validation are done using experimental and independent data.

An important basic operation that has to be followed to successfully handle a problem with Artificial Neural Networks is the selection of a suitable network architecture, which is a choice that mainly depends on the problem.

One of the most frequently employed neural architectures, also in engine applications, is the multi-layer perceptron (MLP). MLP networks consist of successive layers of adaptive weights, with a layer of input neurons, some internal layers (also called *hidden layers*), and a layer of output neurons. The neurons of each layer have connections that run from every unit in one layer to those in the next layers, creating a *feedforward architecture* with no feedback loops (see Fig. 1.2). Therefore, an MLP structure is characterized by the number of neurons of each layer and the number of hidden layers. The number of hidden layer can be increased depending on the problem; however, the most frequent configuration is that with one internal layer only, which is suitable for the majority of the engine modeling and control problems to be handled. On the other hand, big growing networks can be ill-posed for overtraining and be difficult to implement in real time.

Recurrent Neural Networks (RNN) are neural networks with one or more global feedback loops, usually with a unit time delay (often denoted by z^{-1}). The presence of feedback loops introduces a dynamic effect in the computational system and makes them suitable for *black box* nonlinear dynamic modeling and for *input–output mapping*. This feature is of a particular interest in engine applications.

Fig. 1.2 Structure of an MLP neural network model

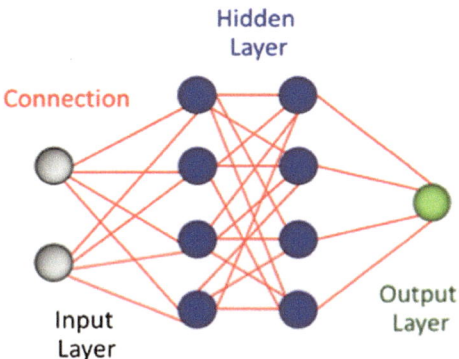

Fig. 1.3 General structure of
a NARX neural network
model

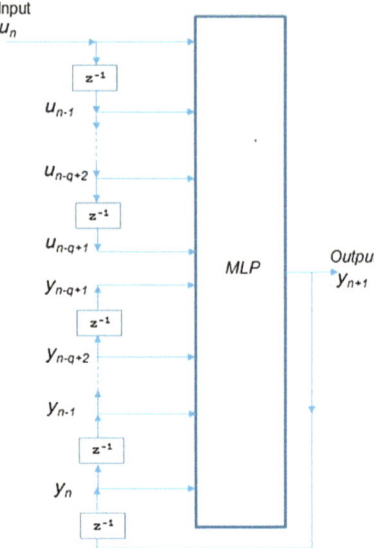

The architectural layout of a recurrent network can take many different forms
mainly dependent on the feedback typology. In nonlinear auto regressive with
exogenous input (NARX) models, the input of a generic multi-layer perceptron is
applied to a tapped delay-line memory of q units, whereas the output is fed back to
the input via another tapped delay-line memory, also of q units. The contents of
these two tapped-delay-line memories are used to feed the input layer of the
multi-layer perceptron [3]. A typical structure of NARX model is shown in Fig. 1.3.

In Fig. 1.3, the present value of the model input is denoted by $u(n)$ and the
corresponding value of the output is denoted by $y(n + 1)$, which is ahead of the
input by one time unit. The formulation of the NARX model can be described as:

$$y(n+1) = F[y(n), \ldots, y(n-q); \ u(n), \ldots, u(n-m+1)], \qquad (1.3)$$

where q is the number of past output terms used to predict the current output, m is
the number of input terms used to predict the current output, and F is a nonlinear
function of *regressors* that are transformations of past inputs and past outputs.
Therefore, NARX neural network structure allows the user to define how many
previous output and input time steps are required for representing the system
dynamics best. This feature can be used in all engine applications where input and
output variables show related dynamics, such as in the case of EGR systems or in
turbocharged engines where turbine and compressor have a strict connection of
process dynamics. However, many other applications, such as Air–Fuel Ratio
(AFR) or exhaust emissions prediction, can be effectively modeled and controlled
using a recurrent network architecture.

A learning algorithm used for recurrent neural networks is the back-propagation
through time algorithm (BPTT), which is a modification of the back-propagation

learning algorithm used for feedforward neural networks. In BPTT training, the network weights are adjusted also on the basis of the network state at previous time steps. For a recurrent neural network trained for a time interval ranging from t_1 to t_n, the total cost function $E(t)$ can be represented as:

$$E(t) = \sum_{t=t_1}^{t_n} E_p(t), \tag{1.4}$$

where the $E(t)$ is the sum of the errors $E_p(t)$ calculated at each time step, and the network weights are adjusted on the basis of the equation:

$$\Delta W_{ij}(t) = -\varepsilon \frac{\partial E(t)}{\partial W_{ij}} = -\varepsilon \sum_{t=t_1}^{t_n} \frac{\partial E_p(t)}{\partial W_{ij}}. \tag{1.5}$$

1.2   Use of ANNs for Modeling and Control of Internal Combustion Engines

The ANNs have been applied to predict the performance of various thermal systems. Their use for modeling the operation of internal combustion engines is more recent. Typically, a neural approach is used to predict the performance and exhaust emissions as well as the specific fuel consumption and fuel–air equivalence ratio of both gasoline and diesel engines [5–9]. For spark ignition engines, ANNs were originally applied to predict the effects of valve-timing on the engine performance and fuel economy [10]. The use of ANN was also proposed to determine torque, brake specific fuel consumption, and emissions in engines using alternative fuels, such as different gasoline–ethanol blends and diesel–biofuel blends [11–13].

In the following paragraphs, some examples concerning the use of neural networks in the modeling and control of some specific engine applications will be presented.

1.2.1   Air–Fuel Ratio Prediction and Control

In order to achieve the optimal functioning of a three-way catalytic converter (TWC), i.e., its maximum efficiency, a spark ignition engine has to operate within a narrow band around the stoichiometric air–fuel ratio (14.7:1), with mean deviations that cannot exceed 0.1% (see Fig. 1.4).

In current technology for gasoline engines, AFR control currently relies on a mean value engine model (MVEM) representation [14, 15]. Such a controller estimates in a feedforward way the actual airflow rate in the cylinder and provides the correspondent amount of fuel to be delivered in the next engine cycle.

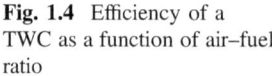

Fig. 1.4 Efficiency of a
TWC as a function of air–fuel
ratio

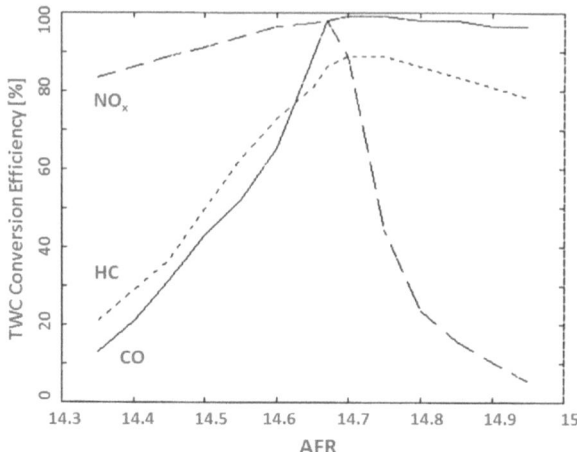

Moreover, the signal of an oxygen sensor placed at the exhaust is used as a feedback control signal to correct the previously calculated mass of fuel to be injected, ensuring a steady state stoichiometry of the mixture.

The fast (but approximate) feedforward component is important to handle transient operating conditions and to compensate the slow dynamics of the feedback loop. There are in fact three time delays included in the value of air–fuel ratio (AFR_o) measured by means of the oxygen sensor: injection delay, combustion delay, and transport delays from the exhaust valve to the oxygen sensor. So:

$$AFR_o(t) = AFR_c(t - t_D), \tag{1.6}$$

where

$$t_D = t_{inj} + t_{comb} + t_{trans}.$$

The transport delay t_{trans} depends on factors such as engine speed, exhaust air mass flow rate, exhaust manifold geometry, etc. [16].

Mean value models for the prediction of airflow rate in the cylinder have some significant limitations, such as the high experimental burden requested for parameters identification and the intrinsic non-adaptive features. To overcome this latter problem, adaptive methodologies (based, for example, on observers, sliding mode controllers, or Kalman filters) have been proposed in order to estimate the states and tune the parameters.

Traditional mean value models for AFR prediction have also to include compensation terms for fuel path dynamics and *wall-wetting* phenomena. In fact, the liquid fuel injected into the intake port only partially enters the cylinder in the current engine cycle. Some of it is collected in fuel films on the walls of intake manifold and close to the back face of the intake valves. The fuel, then, partially evaporates later from these films. A model for this phenomenon was proposed by Aquino [14].

The fuel film dynamics model is highly nonlinear: the fuel fraction and the delay strongly depend nonlinearly on several engine variables (e.g., load, speed, and temperature). Only a good knowledge of these parameters can assure achievement of significant range compensation and hence effective transient control.

The residual gas fraction dynamics together with the mixing dynamics are also highly nonlinear phenomena that have to be modeled for a correct prediction of AFR.

Artificial Neural Networks, which are a powerful tool for modeling highly nonlinear and dynamic systems, can be considered a good candidate for AFR process modeling or for the realization of *virtual* AFR sensors.

Several approaches have been proposed for AFR modeling and control using neural networks. Input parameters to the neural models are generally variables such as engine angular speed, throttle valve opening, absolute manifold pressure, fuel injection time, etc. In [17], AFR was estimated on the basis of spark plug voltage waveform. This latter signal, in fact, is considered to be influenced by the combustion inside the cylinder, which in turn depends, among other factors, on the value of air–fuel ratio. The neural network architecture chosen in this case was a multi-layer perceptron with a cumulative back-propagation learning algorithm. Raw data were firstly pre-processed in order to achieve a satisfactory convergence of the network. To this aim, data points corresponding to regions of the spark signal known to contain poor information about AFR were removed. Moreover, in order to enhance the SNR and reduce the effect of cyclic variation on the spark signal, a filtering was applied to the input data. With regard to the optimal number of training data (input–output vectors pairs), it was stated that a number of training vectors comparable with the number of weights in the net leads to a good generalization capability of the neural network.

In [18], the application of a multiple MLP architecture for adaptive air–fuel ratio (AFR) control is discussed. In this work, the slow AFR process dynamics were represented in the input parameters of the model. In particular, the variable time constant aspects of the process were considered by presenting input parameters to the ANN model as combinations of delayed and filtered sample data values. The filter time constants were designed to span the range of corresponding variable time constants in the system. The approach used to model the variable exhaust transport delay was to configure the ANN model with multiple delayed AFR outputs spanning the full range of speed and load-dependent delays. Of course, this approach is effective if the delays are known or can be calculated. However, when the delays are unknown and variable, a different strategy is required and a further model of the delay itself has to be developed. In this case, the authors propose to train a second supervisory ANN to attribute relevancies to each of the ANN model output predictions. In particular, at high engine speeds, the supervisor attributes more relevance to the shorter time delay model predictions and vice versa.

In [19], neural networks with feedback connections in a recursive architecture were used to model and control the nonlinear air–fuel ratio process dynamics. In particular, the authors used an RNN known in the literature as nonlinear output

Fig. 1.5 General structure of
an NOE model

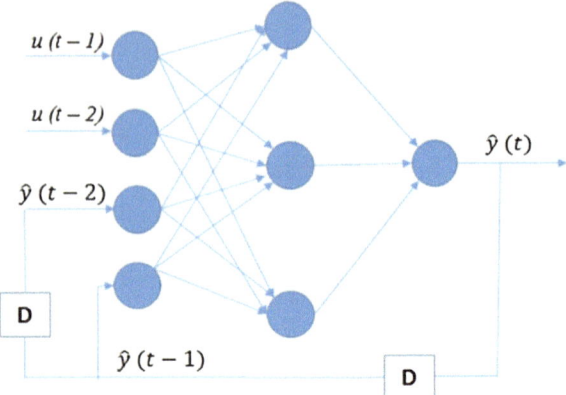

error model (NOE) [20] with one output variable, one hidden layer, and two output
delays, which had a structure like in Fig. 1.5, where the output $\hat{y}(t)$ represents the
actual AFR.

The general form of the NOE model can then be written as:

$$\hat{y}(t|\theta) = F[\hat{y}(t-1|\theta), \hat{y}(t-2|\theta), \ldots, \hat{y}(t-m|\theta), u(t-1), \ldots, u(t-n)], \quad (1.7)$$

where θ is the adjustable parameter, $u(t)$ are reference inputs, the indices m, n define
lag space dimensions of external inputs and feedback variables, and F is a nonlinear
mapping function. The input variables, like in most AFR models, included the
intake manifold absolute pressure (MAP), the engine speed (RPM), and the fuel
injection pulse width (FPW).

The network training is performed by minimizing a cost function estimated as
function of the mean squared error. As for other neural network architectures, the
learning process of a RNN model has to be a compromise between precision and
generalization. High generalization cannot be guaranteed if the training data set
does not provide sufficient information. These latter, in the case of AFR models,
have to cover most of the system operating conditions, providing at the same time a
good knowledge of the dynamic behavior. For this reason, the experimental profiles
of AFR models generally include, other than engine steady state operations, also
sharp accelerations and decelerations. The learning process should also avoid
overtraining that occurs when the minimization task includes many iterations: this
leads to a better precision but to a generalization loss due to overfitting.

A neural controller of injection time based on the information coming from an
AFR neural network model is proposed in [21]. Like in traditional controls, neural
controllers require an identification phase of the process to be controlled that is
known as the *learning control* phase. In case of AFR control, the learning control
phase might use the AFR estimated by a neural model, using the stoichiometric
AFR value and the AFR value measured as output of the engine process by means
of a dedicated sensor.

In order to take into account also process aging effects and to tune the control to the specific engine, the controller configuration can also be done online (online training) during the normal functioning of the engine, even if this could require a high computational effort to the ECU. For this reason, an extremely simple, from a computational point of view, learning algorithm is requested.

Once the controller has been configured, it can commute to the normal operation modality, in which it receives the AFR value measured by the oxygen gas sensor and provides the control variable *injection time*.

The AFR neural controller proposed in [21] had as inputs the engine angular speed, the intake manifold absolute pressure and the angle of throttle opening. A one-step delay was also introduced to take into account all the delays in the measured value of AFR.

In [22], the signal coming from an in-cylinder pressure sensor is used as an input for an injection time control architecture based on a neural AFR virtual sensor. The feedforward part of the control system relies on engine control maps, whereas the feedback part is composed of four subsystems:

- a *features selector* block, which extracts some parameters from pressure curve to be used for AFR prediction;
- a *virtual AFR sensor*, based on an MLP neural network model;
- a block that evaluates both the error and the error variation obtained by comparing the AFR value estimated by the virtual sensor and a reference AFR value (i.e., stoichiometric value);
- a *soft computing controller*, which modifies the injection pulse duration calculated by the feedforward section on the basis of the output of the above block.

The features selector block (see Fig. 1.6) performs a pre-processing of the in-cylinder pressure signal in order to extract the features most relevant to AFR prediction. In other words, it selects the best inputs for the virtual sensor. From an algorithmic point of view, the action of this block is based on *clustering analysis* techniques.

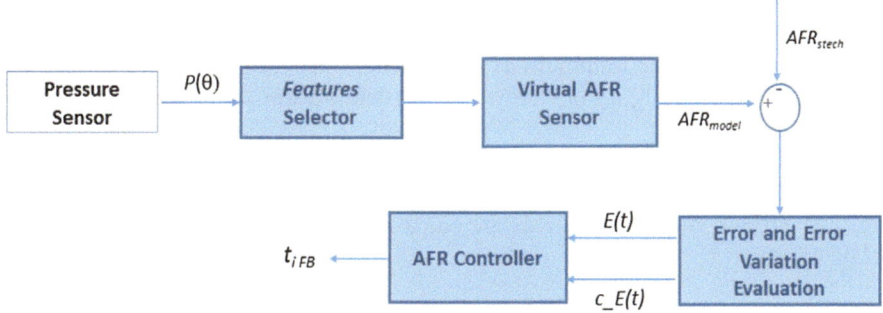

Fig. 1.6 AFR control based on the use of a neural virtual sensor

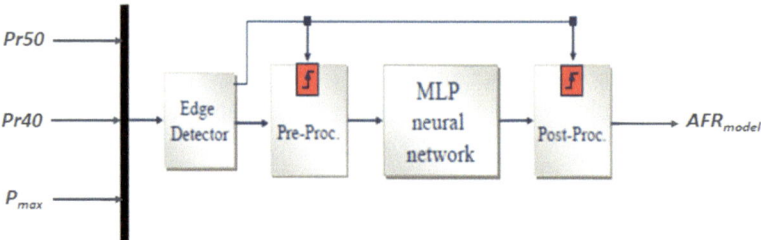

Fig. 1.7 Block scheme of a virtual AFR sensor

Moreover, the block allows to find an optimal sampling of the inputs, i.e., the number of consecutive pressure cycles that have to be averaged to improve AFR prediction.

Figure 1.7 illustrates the block scheme of the virtual AFR sensor. One of the model inputs is *Pr50*, i.e., the combustion pressure value at 50 crank angles (CA) degrees *after* combustion top dead center divided by the value that the same pressure signal assumes at 50 CA *before* top dead center ($Pr50 = P(50)/P(-50)$). Other inputs are *Pr* value at 40 CA ($Pr40 = P(40)/P(-40)$) and the maximum pressure value (P_{max}). These inputs have been chosen, among others, by the *features selector* subsystem.

The *Edge Detector* in Fig. 1.7 is devoted to the synchronization of the model components with the incoming data. It samples the data and provides an enabling signal to the following blocks. The *Pre-Proc* block performs a pre-processing of the incoming data in order to reduce the cycle-by-cycle variation of the input data. To this aim, it carries out a moving average of the input data. The user can modify the number of averaged cycles but, in this case, the optimal value has been chosen by the feature selector.

The heart of the virtual AFR sensor is represented by a multi-layer perceptron neural network. All the endogenous parameters of the neural network, such as the regularization parameter, the number of hidden layers, the number of neurons for each hidden layer, the type of activation function for the neurons, etc., have been set in order to maximize the "generalized forecast capability" of the learning machine. To this aim, a modified version of the *ordinary cross-validation estimate* of the endogenous parameters has been used as fitness function.

The searching of the minimum of this function has been performed by using a stochastic searching algorithm known as particle swarm optimization algorithm (PSOA) (see [23]).

Error and error variation between output of virtual AFR sensor (AFR_{model}) and reference AFR (AFR_{stech}) are calculated at time t in the subsystem shown in Fig. 1.8. To this aim, input signals are suitably dealt with sum blocks and a time delay:

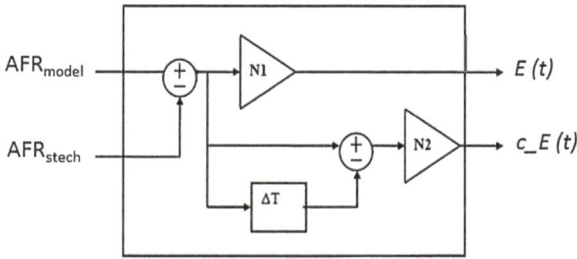

Fig. 1.8 Subsystem for calculation of $E(t)$ and $c\_E(t)$

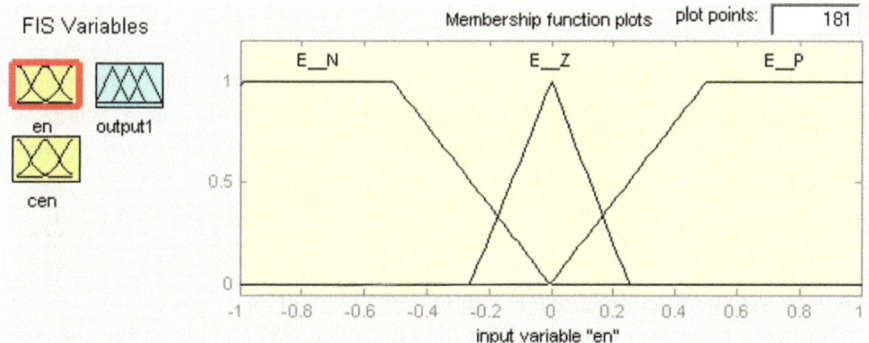

Fig. 1.9 Membership functions of the variable $E(t)$

$$E(t) = N_1 * (\text{AFR}_{\text{stech}} - \text{AFR}_{\text{model}})$$
$$c\_E(t) = N_2 * (E(t) - E(t - \Delta T));$$

where N_1 and N_2 are two normalization factors.

The variables $E(t)$ and $c\_E(t)$ represent the input to the AFR control system. AFR soft computing control was implemented by means of fuzzy techniques. In particular, the following fuzzy sets of the input variable $E(t)$ were considered:

- E_P "*positive error,*"
- E_N "*negative error,*"
- E_Z "*zero error,*"

with the corresponding membership functions shown in Fig. 1.9.
The following fuzzy sets have been considered for the variable $c\_E(t)$:

- CE_P "*variation of the positive error*";
- CE_N "*variation of the negative error*";
- CE_Z "*variation of the zero error*";

with the corresponding membership functions depicted in Fig. 1.10.

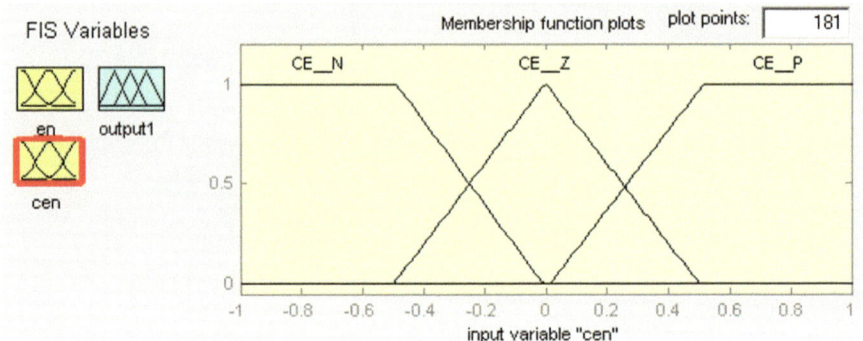

Fig. 1.10 Membership functions of the variable C_*E*(*t*)

The fuzzy controller is a fuzzy system having as antecedents the fuzzy sets of *E* (*t*) (error) and *c_E*(*t*) (error variation), and as consequents the fuzzy sets of the output variable Δ_DI (injection time).

- I_N "*positive injection,*"
- I_P "*negative injection,*"
- I_Z "*zero injection,*"

whose membership functions are shown in Fig. 1.11.

The fuzzy algorithm conceived for the controller was composed by nine rules having the following traditional form:

- *IF E_P AND CE_P THEN output IS I_P*
- *IF E_N AND CE_N THEN output IS I_N*
- *IF E_P AND CE_Z THEN output IS I_P*
- *IF E_N AND CE_Z THEN output IS I_N*
- *IF E_P AND CE_N THEN output IS I_Z*
- *IF E_N AND CE_P THEN output IS I_Z*

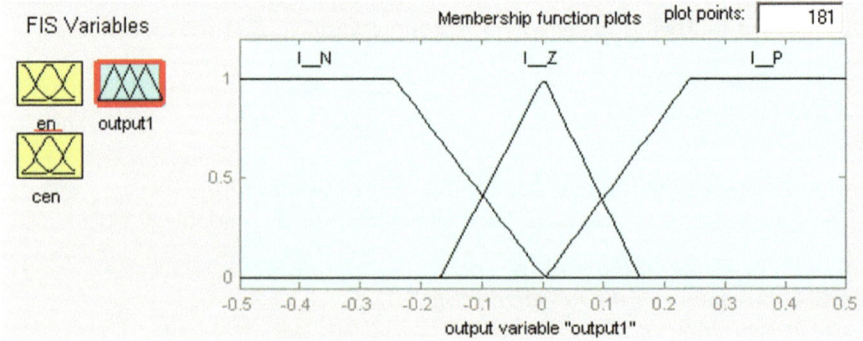

Fig. 1.11 Membership functions of the output variable Δ_DI

Fig. 1.12 Tridimensional map of AFR fuzzy controller

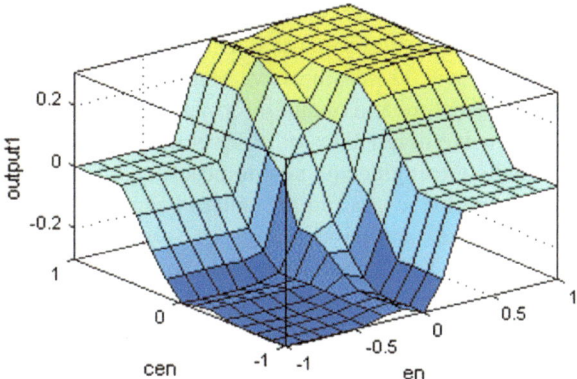

- *IF E_Z AND CE_Z THEN output IS I_Z*
- *IF E_Z AND CE_N THEN output IS I_N*
- *IF E_Z AND CE_P THEN output IS I_P*

The aim of the fuzzy algorithm described by the previous rules is to provide for each value of input variables a correspondent value of the control variable Δ_DI (injection time). Figure 1.12 illustrates the tridimensional map summarizing the way of functioning of the controller.

Finally, to estimate the update of the control variable induced by the feedback part of the air–fuel control system the output of the controller is processed in the following way:

$$t_{i\,FB} = N_3 * \Delta\_DI(t) + \Delta\_DI(t - \Delta T),$$

where N_3 and ΔT are two factors to be tuned.

The proposed approach allowed to obtain a strict control of AFR with an error, with an error below of 1% both in steady state and during fast transients.

1.2.2 Use of Neural Networks to Predict Combustion Pressure Parameters

Real-time combustion process monitoring in internal combustion engines may provide a strong tool regarding engine operation and may be profitably used for closed-loop electronic engine controls that allow internal combustion engines to comply with the severe normative on pollutants emission and fuel consumption. One of the most important parameters used for the evaluation of the combustion quality is the in-cylinder pressure. However, this kind of measure requires an intrusive approach to the cylinder and a special mounting process. Moreover, the combustion pressure transducers used for this kind of applications still have a high cost for mass production automotive engines and still remains some problems of robustness and performances. Due to the disadvantage of direct pressure measurement, several non-intrusive techniques have been proposed to reconstruct the cylinder pressure and obtain information about the combustion quality [24–26]. The non-intrusive diagnostics offer several advantages: the sensors are generally placed externally the engine and no engine structural modifications are required. Moreover, the non-intrusive sensors are quite cheap, as they are not requested to resist very high pressures and temperatures.

Artificial Neural Networks can have a role in estimating several combustion pressure parameters on the basis of information coming from sensors already present on engine. An example of this is combustion pressure monitoring using the *engine crankshaft speed*.

Cylinder pressure reconstruction on the basis of instantaneous crankshaft speed is considered to be a successful approach both for its simplicity and its low cost [27–32]. Crankshaft angular speed, in fact, contains information to reconstruct cylinder pressure, and hence, it can be chosen as an input to a neural network model for cylinder pressure estimation. In particular, the following correlation can be written:

$$\dot{\omega} = \frac{1}{J(\theta)} \left[-\frac{1}{2} \frac{dJ(\theta)}{d\theta} \omega^2 + T_{\text{ind}} - T_{\text{fric}} - T_{\text{load}} \right], \tag{1.8}$$

where ω is the crankshaft velocity, $J(\theta)$ is the moment of inertia of the engine, T_{ind}, T_{fric}, and T_{load} are the indicated torque, frictional torque, and load torque, respectively. The indicated torque is strictly correlated to the in-cylinder pressure $P(\theta)$, and for a single cylinder engine, it is given by:

$$T_{\text{ind}} = (P(\theta) - P_{\text{man}})A_{\text{p}} \frac{ds(\theta)}{d\theta}, \tag{1.9}$$

where A_{p} is the piston head surface area, P_{man} is the intake manifold pressure, and s (θ) is the piston stroke from top dead center (TDC). Therefore, Eq. (1.8) can be re-written as:

$$\dot{\omega} = f(\theta, P, \omega) = \frac{1}{J(\theta)} \left[-\frac{1}{2} \frac{dJ(\theta)}{d\theta} \omega^2 + (P(\theta) - P_{\text{man}})A_{\text{p}} \frac{ds(\theta)}{d\theta} - T_{\text{fric}} - T_{\text{load}} \right]. \tag{1.10}$$

Equation (1.10) illustrates clearly the nonlinear correlation existing between engine crankshaft speed and in-cylinder pressure.

In [31], cylinder pressure reconstruction was carried out using a RNN with instantaneous crankshaft speed fluctuations and motored pressure as input signals. In addition, spark advance was chosen as another input, since the values of peak pressure and its location and the rate of pressure rise are significantly affected by the ignition time value. Equivalence ratio also has a significant influence on the cylinder pressure peak, hence it was chosen as one of the inputs. On the basis of these inputs, the regression vector for neural network was given by:

$$P(t|w) = NN(P(t-1), \ldots, P(t-a), \omega(t-1), \ldots, \omega(t-b), P_{mot}(t-1),$$
$$P_{mot}(t-c), SA(t-1), \varnothing(t-1), w) \tag{1.11}$$

where a, b, and c are the number of past cylinder pressure output, the number of past speed, and past motored pressure data given to the neural network as feedback variables.

The authors demonstrated that the minimum mean square error in pressure prediction was obtained for $a = 3$, $b = 7$, $c = 7$, and with 13 neurons in the hidden layer. By using the above-described RNN model, the residuals between the experimental pressure curve and the simulated one lay in the range ±0.5 bar for all the investigated operating conditions.

In [32], a radial basis function (RBF) neural network model having as input the instantaneous angular speed was used for pressure waveform reconstruction. Using an RBF, the correlation between cylinder pressure P and engine angular speed is given by:

$$P_{kj}(\theta) = \sum_{i=1}^{n} h_{ki}(\theta) w_{ij}.$$

Therefore, the pressure can be expressed as a linear combination of the output of the hidden neurons, that is, as a linear combination of a set of n fixed basis functions $h_{ki}(\cdot)$; the coefficients of the linear combinations, w_{ij}, are the weights or model parameters.

The results of the proposed approach showed that the measured and estimated pressure traces matched well over all stages of the pressure process: compression, onset of combustion, peak pressure, and the rise and fall of the combustion.

The maximum deviation interval between the predicted and measured pressures resulted less than 0.46 bar. This is the same quantitative order as that of the cycle-to-cycle variation and, therefore, the prediction can be judged to be accurate enough for averaged analysis over the cyclic variation. Deviation of peak pressure was less than 3 bar, whereas the deviation of the angular location of pressure peak was ±1 crank angle.

In [33], a neural network approach for real-time prediction of in-cylinder pressure peak value (PP) and its angular location (LPP) has been proposed. The trained network, which can be viewed as a nonparametric model of the engine process, had as inputs the engine angular crankshaft speed and the crankshaft speed derivative (i.e., crankshaft acceleration) (see Fig. 1.13).

In order to train and validate the neural network model, measurements were carried out over the engine speed range 1000–2000 rpm, with steps of 200 rpm, and absolute intake pressure values ranging from 1000 to 1600 mbar.

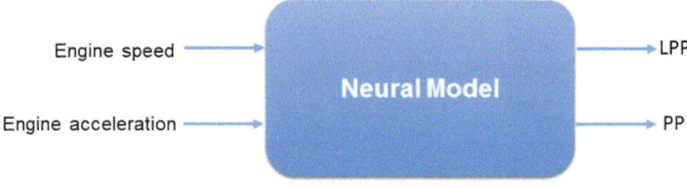

Fig. 1.13 Neural network model for prediction of LP and LPP

Table 1.1 Main features of the neural model for PP and LPP prediction

Neural network structure	feedforward MLP
Neuron model	tan$h(x)$
# Neurons in hidden layer	30
Training algorithm	*Trainbr* with 0.3 as regularization factor

The overall data set was, then, divided into two groups. One group was used as training data set, i.e., to train the neural network model and to set the internal model parameters. The other group was used as testing data set, i.e., to validate the trained network.

As neural model, a multi-layer perceptron (i.e., MLP) neural network was chosen. The MLP had only a hidden layer with 30 neurons and *arctan* as activation function. In order to train the network, a Bayesian regularization back-propagation was used. This latter is a process that minimizes a combination of squared errors and weights and then determines the correct combination so as to produce a network that generalizes well. Use of this process guaranteed a satisfactory generalization capability of the neural network model and, at the same time, allowed to avoid overfitting issues. The tuning of internal parameters (e.g., the value of the regularization parameter, neuron biases) of the neural network model was optimized by using an evolutionary algorithm. In Table 1.1, the main features of neural network model are shown.

The overall *Relative Error* in the prediction of pressure peak (PP), obtained from scaling the root mean square error (RMSE) by the maximum of the peak pressure of the experimental data set, was 4% at 1000 rpm, 5% at 1500 rpm, and 7% at 2000 rpm. The model also showed a good capability of predicting the angular location of pressure peak (LPP) with a RMSE ranging from 1.38 to 5.2 crank angles degrees.

The results confirmed that the neural model can be effectively used for PP and LPP estimation. Moreover, the model also revealed its capability to predict pressure peak reductions due for example to inefficient combustions, misfiring events, and other combustion abnormalities. As a consequence, it can be used as a non-intrusive tool for real-time diagnosis of engine combustion quality in advanced closed-loop control systems.

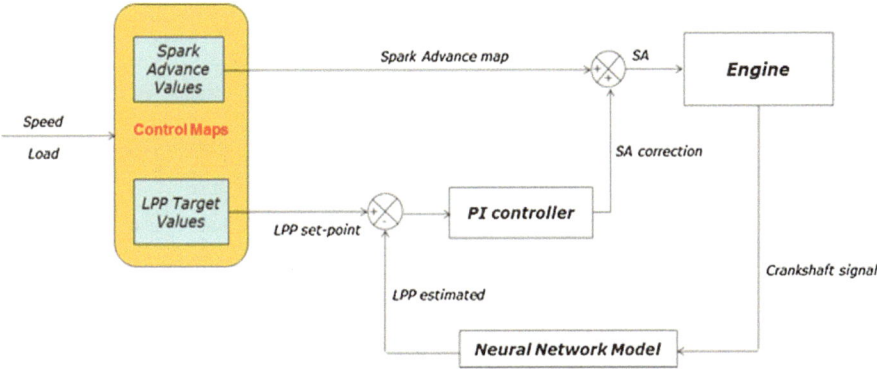

Fig. 1.14 Spark advance closed-loop control based on LPP prediction by means of a neural model

Location of pressure peak, as it is strictly connected to the ignition angle, represents a promising control variable for gasoline engine spark advance control. A possible engine controller could estimate the LPP from the crankshaft speed information and could use it as a feedback variable in an ignition timing controller. This allows to maintain the LPP close to its reference value modifying, if requested, the spark advance value stored in the engine control maps. For each engine speed and engine load, the LPP set-point can be defined as the optimal value to obtain the desired engine behavior. In high load ranges, for example, late pressure peak locations could be requested to hold down the NOx emissions.

A block scheme of a possible engine closed-loop control is shown in Fig. 1.14. The architecture includes a proportional–integral controller for LPP control.

A similar controller was proposed in [34]. Also in this work, a spark advance control strategy based on the location of peak pressure (LPP) is presented. A feedforward MLP neural network is introduced in this study to predict LPP by using only few samples (five) extracted from in-cylinder pressure voltage signal (the entire acquisition of pressure curve was not required).

References

1. J.M. Zurada, *Introduction to Artificial Neural Systems* (West Publishing Company, St. Paul, MN, 1992)
2. F. Rosemblatt, The perceptron: a probabilistic model for information storage and organization in the brain. Psychol. Rev. **65**, 386–408 (1958)
3. S.S. Haykin, *Neural Networks and Learning Machines* (Prentice Hall, 1994)
4. L. Fortuna, G. Rizzotto, M. Lavorgna, G. Nunnari, G. Xibilia, R. Caponetto, *Soft Computing— New Trends and Applications* (Springer, Heidelberg, 2001)
5. S.A. Kalogirou, Application of artificial neural-networks for energy systems. Appl. Energy **67**, 17–35 (2000)

6. C. Sayin, H.M. Ertunc, M. Hosoz, I. Kilicaslan, M. Canakci, Performance and exhaust emissions of a gasoline engine using artificial neural network. Appl. Therm. Eng. **27**, 46–54 (2007)
7. E. Arcaklioglu, I. Celikten, A diesel engine's performance and exhaust emissions. Appl. Energy **80**, 11–22 (2005)
8. G. Najafi, B. Ghobadian, T. Yusaf, H. Rahimi, Combustion analysis of a CI engine performance using waste cooking biodiesel fuel with an artificial neural network. Am. J. Appl. Sci. **4**(10), 756–764 (2007)
9. N. Ladommatos, R. Balian, R. Horrocks, L. Cooper, The effect of exhaust gas recirculation on soot formation in a high-speed direct-injection diesel engine. SAE Technical Paper 960841
10. M. Golcu, Y. Sekmen, P. Erduranli, S. Salman, Artificial neural network based modelling of variable valve-timing in a spark ignition engine. Appl. Energy **81**, 187–197 (2005)
11. G. Najafi, B. Ghobadian, T. Tavakoli, D.R. Buttsworth, T.F. Yusaf, M. Faizollahnejad, Performance and exhaust emissions of a gasoline engine with ethanol blended gasoline fuels using artificial neural network. Appl. Energy **86**, 630–639 (2009)
12. B. Ghobadian, H. Rahimi, A.M. Nikbakht, G. Najafi, T.F. Yusaf, Diesel engine performance and exhaust emission analysis using waste cooking biodiesel fuel with an artificial neural network. Renew. Energy **34**, 976–982 (2009)
13. M.I. Canakc, A. Erdil, E. Arcaklioglu, Performance and exhaust emissions of a biodiesel engine. Appl. Energy **83**, 594–605 (2006)
14. C.F. Aquino, Transient A/F control characteristics of the 5 liter central fuel injection engine. SAE Paper 810494 (1981)
15. E. Hendricks, S. Sorenson, SI engine controls and mean value engine modelling. SAE Technical Paper 910258 (1991). doi:10.4271/910258
16. A. di Gaeta, S. Santini, L. Glielmo, F. De Cristofaro et al., An algorithm for the calibration of wall-wetting model parameters. SAE Technical Paper 2003-01-1054 (2003). doi:10.4271/2003-01-1054
17. J. Howlett, S.D. Walters, P.A. Howson, I.A. Park, Air-fuel ratio measurement in an internal combustion engine using a neural network, in *Advances in Vehicle Control and Safety International Conference, AVCS*, vol. 98 (1998)
18. A.M. Frith, C.R. Gent, A.J. Beaumont, Adaptive control of gasoline engine air-fuel ratio using artificial neural networks, in *4th International Conference on Artificial Neural Networks* (1995), pp. 274–278
19. I. Arsie, M. Sorrentino, C. Pianese, A neural network air-fuel ratio estimator for control and diagnostics in spark-ignited engines. IFAC Proc. Vol. **40**(10), 227–234 (2007)
20. M. Nørgaard, O. Ravn, N.L. Poulsen, L.K. Hansen, *Neural Networks for Modelling and Control of Dynamic Systems* (Springer, Heidelberg, 2000)
21. C. Alippi, C. de Russis, V. Piuri, A neural-network based control solution to air-fuel ratio control for automotive fuel-injection systems. IEEE Trans. Syst. Man Cybern. Part C (Appl. Rev.) **33**(2) 259–268 (2003)
22. N. Cesario, M. Di Meglio, F. Pirozzi, G. Moselli et al., Air/fuel control system in SI engines based on virtual lambda sensor. SAE Technical Paper 2005-24-058 (2005). doi:10.4271/2005-24-058
23. V.N. Vapnik, An overview of statistical learning theory. IEEE Trans. Neural Netw. **10**(5), 988–999 (1999)
24. C. Mobley, Non-intrusive in-cylinder pressure measurement of internal combustion engines. SAE Paper no 1999-01-0544 (1999)
25. D. Panousakis, J. Patterson, A. Gasiz, R. Chen, Analysis of SI combustion diagnostics methods using ion-current sensing techniques. SAE Paper no 2006-01-1345 (2006)
26. M. Wlodarczyk, High accuracy glow plug-integrated cylinder pressure sensor for closed loop engine control. SAE Paper no 2006-01-0184 (2006)
27. S.J. Citron, J.E. O'Higgins, L.Y. Chen, Cylinder by cylinder engine pressure and pressure torque waveform determination utilizing speed fluctuations. SAE Paper no 890486 (1989)

28. G. Rizzoni, Diagnosis-of individual cylinder misfires by signature analysis of crankshaft speed fluctuations. SAE Paper no 890884 (1989)

29. F.T. Connolly, A.E. Yagle, Modeling and identification of the combustion pressure process in internal combustion engines using engine speed fluctuations. Am. Soc. Mech. Eng. Dyn. Syst. Control Div. **44**, 191–206 (1992)

30. D. Moro, N. Cavina, F. Ponti, In-cylinder pressure reconstruction based on instantaneous engine speed signal. J. Eng. Gas Turbines Power **124**, 220–225 (2002)

31. S. Saraswati, S. Chand, Reconstruction of cylinder pressure for SI engine using recurrent neural network. Neural Comput. Appl. **19**, 935–944 (2010)

32. F. Gu, P.J. Jacob, A.D. Ball, A RBF neural network model for cylinder pressure reconstruction in internal combustion engines, in *IEEE Colloquium on Modelling and Signal Processing for Fault Diagnosis (Digest no: 1996/260)* (1996)

33. F. Taglialatela, M. Lavorgna, E. Mancaruso, B.M. Vaglieco, Determination of combustion parameters using engine crankshaft speed. Mech. Syst. Signal Process. **38**(2), 628–633 (2013)

34. S. Park, P. Yoon, M. Sunwoo, Feedback error learning neural networks for spark advance control using cylinder pressure. Proc. Inst. Mech. Eng. Part D: J. Automob. Eng. **215**(5), 625–636 (2015)

Chapter 2
Non-interfering Diagnostics for the Study of Thermo-Fluid Dynamic Processes

The conversion of chemical energy into mechanical power, operated by internal combustion engines, involves a great number of complex phenomena that often occur in transient thermo-fluid dynamic conditions. The majority of these phenomena are affected by nonlinear dynamics, thus requiring appropriate compensation techniques.

The analysis and comprehension of these nonlinear processes is a basic requirement for the design of effective control solutions, able to optimize the combustion processes in terms of engine power, efficiency, and emissions.

In this chapter, we present some advanced non-interfering optical diagnostics that allow to study in detail the reasons and the effects of the nonlinear behavior of many processes occurring in internal combustion engines.

The internal combustion (IC) engine is designed to provide power from the energy that fuel contains. More specifically, the chemical energy of fuel and air creates a mixture that burns to produce mechanical power. There are various types of fuels that can be used in IC engines, which include liquid (gasoline, Diesel fuel, biofuels) and gaseous (methane, propane, hydrogen) substances. The output power produced by an IC engine results from the fuel that it uses and its mechanical parts. In an internal combustion engine, a piston moves up and down in a cylinder, and the power produced is transferred through a connecting rod to a crankshaft. Most engines operate on what is known as four-stroke cycle. Each piston completes four separate strokes while turning a crankshaft. A stroke refers to the full travel of the piston along the cylinder, in either direction [1]. Both spark ignition (SI) and compression ignition (CI) engines use this cycle, which comprises:

- Intake stroke, in which fresh air enters and the fuel can start to spray;
- Compression stroke, in which the fluid and fuel are compressed and, at end, the combustion can occur;
- Expansion stroke or power stroke, when the combustion evolves and exhaust gases, including pollutants, are produced;
- Exhaust stroke, when the exhaust gases are emitted.

© The Author(s) 2018
F. Taglialatela Scafati et al., *Nonlinear Systems and Circuits in Internal Combustion Engines*, SpringerBriefs in Nonlinear Circuits,
https://doi.org/10.1007/978-3-319-67140-6_2

The processes occurring during these cycles consist of a complex interaction of homogenous and heterogeneous chemistry and transport processes. The processes involved in internal combustion engines (ICE) are two-phase, turbulent mixing-controlled, and they include short timescale phenomena such as turbulence production and dissipation, spray breakup and evaporation, and pollutants formation.

As consequence, in order to carry out appropriate investigations, non-intrusive diagnostic techniques with high temporal and spatial resolution must be considered. In these years, the science of optics has played an important role in the measurement and understanding of combustion phenomena, including not only laboratory flames but also practical devices such as the internal combustion engines. Optical diagnostics allows to obtain pertinent knowledge in complex systems where nonlinear and very fast phenomena occur. In particular, the use of optical engines with wide accessibility, improved detectors, and ultra-fast light sources tools is considered of great interest, as it allows real-time and high spatial resolution characterization of the phenomena without any modification of the different phases occurring.

There are numerous methods applied in optical engines, and just a few of them have the potential of being applied in practical engines. Optical diagnostics operate in a broad spectral range, from ultraviolet (200 nm) to infrared (10,000 nm). In each spectral domain, it can apply diagnostics based on emission, absorption, and scattering fluorescence measurements. The decision for application of a specific optical technique is always based on the requested information other than the efforts and chances to gain the required results with a given method.

The present chapter provides an overview of optical methods and their application to the study of the complex phenomena occurring in internal combustion engines.

2.1 Air Motion

The study of fluid mechanics in the cylinder in terms of mean motion and turbulence characteristics can be made by means of Laser Doppler Anemometry (LDA), also known as Laser Doppler Velocimetry (LDV), Particle Image Velocimetry (PIV), and Phase Doppler Anemometry (PDA). By these techniques, it is possible to analyze the air motion, which contributes both to the air/fuel mixing and to the flame propagation process during combustion. The LDA is a technique that uses the Doppler shift in a laser beam to measure the velocity [2]. It is an optical method, tightly related to both the physical and geometrical optics, and it is a widely accepted tool for investigating fluid dynamics with gases and liquids. As the fluid moves through the probe volume with small seeding particles, the information of the flow velocity is generated. In particular, when a particle passes through the intersection volume (measurement volume) formed by the two coherent laser beams, the scattered light, received by a detector, has components from both beams.

The intensity of the scattered light has a fundamental frequency that is connected to the particle velocity.

LDA usually provides single-point velocity statistics, typically in two directions, while PIV provides instantaneous velocity images. LDA can be arranged to measure local flow velocity, even the smallest length scales of the flow (the Kolmogorov scale), and it provides single-point statistics, which are well matched to time-averaged modeling approaches (e.g., RANS models). While it is usually quite difficult to acquire PIV data down to the Kolmogorov scale, PIV provides additional insight by allowing researchers to visualize an entire field and thus to visualize phenomena like spatial correlations that are well matched to LES models. For reasonably laminar or transitional flows, PIV is often used by itself. For turbulent flows as air motion in the ICE, the problem becomes much more challenging. LDA and PIV are often used in complementary ways for such flows. Both LDA and PIV require that the flow be seeded with particles, and this seeding process is usually different from one flow to the next. To apply PIV to new environments usually requires the researcher to evaluate the needed seed density for good spatial resolution. During the combustion, this technique lacks in sensibility: very high seed density can affect flame temperature by enhancing radiation heat transfer from the flame. One must also be careful to consider factors like inter-pulse spacing to produce best quality correlation of the two PIV images. In addition, three-dimensional flows require one to consider out-of-plane motion and its effect on accuracy. To apply PIV to challenging environments can require special arrangements, such as a fast shutter to avoid high background levels caused by flame radiation.

2.2 Mixture Preparation

Mixture preparation is of major importance when designing next generation internal combustion engines. To minimize engine out emissions, such as CO_2, CO, NO_x, and particulate matter, while enhancing performance and fuel economy, it is crucial to have control over this parameter. The investigation of the fuel and air mix within the cylinder can be carried out by several techniques that have been widely used in the past and now greatly improved by new intensified charge-coupled device (ICCD) cameras, complementary metal-oxide semiconductor (CMOS) cameras, and less expensive laser sources.

During the last decade, various laser techniques have clearly shown their effectiveness in measurement and diagnostics of various combustion phenomena. These techniques have the great benefit of being able to make non-intrusive measurements (no physical probe introduced in the combustion zone), to have high spatial (<0.001 mm^3) and temporal resolution (<10 ns), and they can measure several points/species in one laser pulse. In addition, many techniques can provide two- or even three-dimensional information.

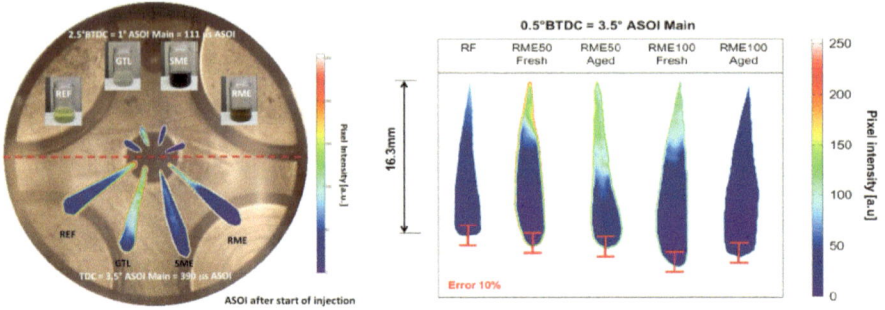

Fig. 2.1 Characteristics of different fuels [commercial Diesel (REF), GTL and several biofules (SME, RME blends and RME aged)] obtained by luminous intensity distributions in the jet

They are based on the high-speed direct, backlight, and Schlieren imaging [3] and can be applied to study spray atomization, the fuel penetration and dispersion, as well as the evaporation phenomena. Although cycle resolved, these techniques cannot provide spatial quantitative information on the air–fuel mixing in the combustion chamber, but only statistical data can be obtained, owing to the instability and highly turbulent nature of Diesel sprays and of high injection pressure of new gasoline direct injection spray. Nevertheless, the liquid tip penetration and cone angle can be obtained with good accuracy for the main injection, meanwhile less information can be obtained for short pre- and pilot injections. Recently, interesting results about the pre- or pilot injection were obtained with use of infrared diagnostics. In [4], the visualization of the pilot injection process was obtained by means of an optical access in the piston head and by the presence of an inclined mirror for the collection of the images. The pilot injection was characterized by fast IR imaging. This innovative and non-conventional optical technique in the infrared was applied for the visualization of the liquid and vapor fuel. Other authors [5] performed in the infrared (IR) region of the spectrum black-body emission an imaging of the temperature reduction caused by the impingement of the spray on an electrically heated wall. However, the metallic (or fused silica) surfaces in the cylinder cannot be treated as a black body. Hydrocarbon fuels absorb IR light over broad wavelength regions. By measuring the attenuation of transmitted IR light, it would be possible to detect even thin wall films. A disadvantage is that optical access to the combustion chamber from two opposing sides is needed for a transmission measurement.

Developments in laser source and detection systems allowed to set new 2-D laser imaging diagnostics to obtain the spatial distribution of liquid and vapor of fuel inside optically accessible engines by simultaneous use of laser-induced Rayleigh and Mie scattering imaging, and EXCIPLEX technique, based on a fluorescence system. This latter technique, even if it can distinguish simultaneously between the spray liquid and vapor phases, needs of a fluorescence dopant, and therefore, it is limited to oxygen-free atmosphere because the doping molecule is severely quenched by the oxygen and a surrogate fuel must be used [2].

A different technique, based on the principle of absorption of ultraviolet laser light by vapor fuel and the scattering of visible laser light by liquid fuel, seemed to give good results because it measures simultaneously the concentrations of vapor and liquid in an evaporating Diesel spray. Being an absorption technique, its drawback is the integration across the entire width of spray. The technique can be used in optical engines using a polychromatic light beam, and it is based on Lambert–Beer's law. Considering that a light beam of an intensity I_0 is attenuated when it passes through a multi-phase spray, the transmitted light intensity I_t can be written as follows:

$$\frac{I_t}{I_0} = \exp(-K_{\text{ext}} \cdot L) \tag{2.1}$$

where L is the optical path length and K_{ext} the extinction coefficient that is equal to:

$$K_{\text{ext}} = -\frac{\pi}{4} C_n \int Q_{\text{ext}} N(D) D^2 \mathrm{d}D \tag{2.2}$$

where C_n is the number density of droplets, Q_{ext} is the extinction efficiency, D is the droplet diameter, and $N(D)$ is the droplet size distribution function [2]. In the UV and visible range, the spectral extinction is due to different phenomena which can be separately analyzed [6]. Using the principle of absorption of ultraviolet light by fuel vapor and the scattering of light in the visible range by liquid fuel droplets, the concentration of both phases can be obtained. The visible light, which is not absorbed by both liquid and vapor phases, is attenuated only by the scattering caused by droplets. Therefore, the transmissivity of visible light by the fuel spray is equal to the transmissivity due only to the scattering caused by liquid droplets, as shown by the following equation:

$$\textbf{Visible}: \quad \log\left(\frac{I_0}{I_t}\right)_{\text{Vis}} = \log\left(\frac{I_0}{I_t}\right)_{L,\text{sca}} \tag{2.3}$$

The ultraviolet light, which is absorbed by both liquid and vapor phases, is attenuated due to the absorption caused by vapor, to the scattering caused by liquid droplets, and to the absorption of liquid droplets according to the equation:

$$\textbf{UV}: \quad \log\left(\frac{I_0}{I_t}\right)_{\text{UV}} = \log\left(\frac{I_0}{I_t}\right)_{V,\text{abs}} + \log\left(\frac{I_0}{I_t}\right)_{L,\text{sca}} + \log\left(\frac{I_0}{I_t}\right)_{L,\text{abs}} \tag{2.4}$$

The last two terms are much lower than the first one, so the UV transmissivity depends mainly on the absorption by fuel vapor [6].

Considering Eqs. (2.1) and (2.2) and experimental extinction data in the visible reported in the hypothesis of Eq. (2.3), it can be obtained:

$$\log\left(\frac{I_0}{I_t}\right)_{vis} = \left(-\frac{\pi}{4}C_n \int Q_{ext}N(D)D^2\mathrm{d}D\right)L \qquad (2.5)$$

From this equation, the mass concentration of fuel droplets per unit volume, C_d, considering the results of [6], can be expressed as:

$$C_d = \frac{2}{3L}\rho_f\frac{D_{32}}{Q_m}\log\left(\frac{I_0}{I_t}\right)_{Vis} \qquad (2.6)$$

where ρ_f is the liquid fuel density, Q_m is the mean extinction efficiency for the poly-dispersed particle cloud, and D_{32} is known as the volume/surface mean diameter (Sauter diameter), which can be obtained by Phase Doppler Particle Analyzer (PDPA) measurements or by the correlation of Hiroyasu [7]. Analogous expression can be obtained for the vapor concentration, C_v, considering Eqs. (2.1) and (2.4), as reported in [6]:

$$C_v = \frac{1}{kL}\log\left(\frac{I_0}{I_t}\right)_{V,abs} \qquad (2.7)$$

where k is the molar absorptivity of the vapor fuel.

A comparison between liquid and vapor distribution for n-heptane and Diesel fuel, obtained from extinction measurements at different times, after Start of Injection, can be found in [8], where the measurements are carried out in an optical Diesel engine using a commercial Diesel fuel and n-heptane.

In order to better understand the liquid distribution of the different fuels, the injection images detected in the combustion chamber can be processed and converted on 8-bit scale. A scale of 256 colors is used to represent the several luminous levels into the jet (Fig. 2.1). The RME fuels show a denser liquid core with respect to the reference and a higher penetration because of the high density/viscosity.

2.3 Combustion

In the past, optical analysis of the intermediate steps from ignition to soot formation was generally restricted to the measurements of ignition delay, and the high-speed imaging was performed doping the fuel with copper to create a more luminous emission before the formation of soot. The lack of spatial and time resolution typical of these techniques has been partially overcome with the natural chemiluminescence imaging by using ICCD and CMOS cameras.

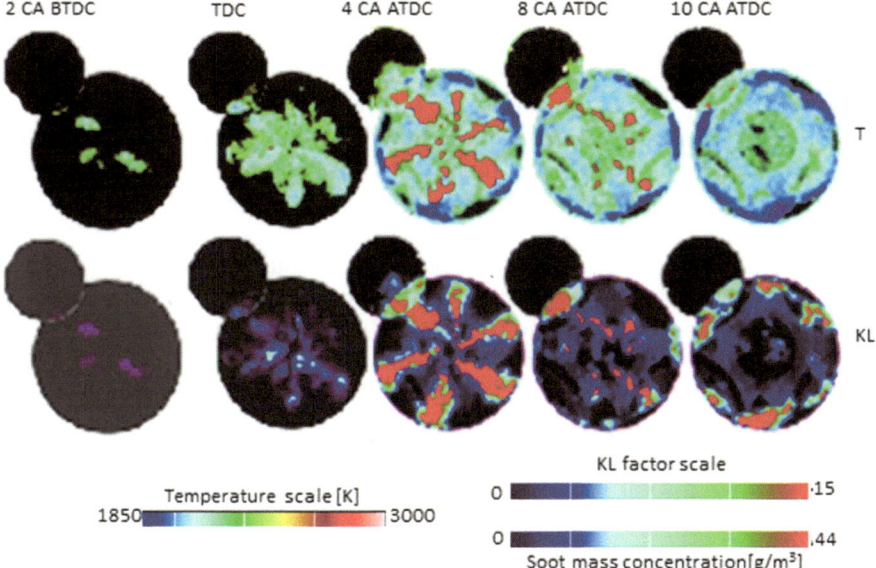

Fig. 2.2 Temperature and soot evolution in a Diesel engine

In [9], the combustion evolution in an optical common rail Diesel engine detected by chemiluminescence measurements using a visible narrow filter on a CCD camera is reported. By this diagnostics, useful information on turbulent evolution of flame in the combustion chamber and on the species involved in this spectral range can be obtained. Radiation of a Diesel flame is essentially due to thermal radiation of the soot particles formed under heterogeneous diffusion combustion of the fuel sprays. Its visible intensity is used to evaluate temperature and soot concentration comparing it with Planck's radiation formula. This "two color" method was applied to obtain the results presented in Fig. 2.2 for the same conditions reported in [9]. It is based on the use of two selected wavelengths (colors) in the spectrum, which are used to obtain, with an iterative procedure, the absolute temperature by Planck's formula.

Analogous results can be obtained for SI engines, even if the soot produced is very low. In order to overcome this limitation and to consider the effect of cycle-to-cycle variation typical of SI engine, a CMOS camera is used in [10]. In this work, a selection of images detected during the combustion phase of ethanol and gasoline for injection pressures of 30 and 120 bar is shown. The images were obtained through an optical window on a piston of an optical accessible spark ignition (SI) engine. For both fuels, a luminous arc represents the evidence of spark ignition. It occurs at around 24° CA BTDC and persists until 15° CA BTDC, when the flame kernel is well observable, even if its luminosity is much lower than the spark. Then, the flame kernel moves from the spark plug in the combustion chamber, and it is the highest for the gasoline injected at 30 bar, whereas it decreases when the injection pressure increases, due to a better fuel–air interaction. About ethanol, the soot amount production increases at the rise of injection pressure due to the amount of fuel impinged on the piston. In this case, it is important to note that at the opening of exhaust valves, the differences of soot amount in the combustion chamber decrease. Moreover, the development of the combustion process can be analyzed by means of a post-processing of the optical data. In particular, the flame front speed and radius, which are key parameters in the study of combustion process in spark ignition engines, can be evaluated.

These techniques also allow the detection of anomalous combustion occurring in SI engine known as knock. Knock is a consequence of a spontaneous ignition of a portion of the fresh mixture ahead the propagating flame. It gives the name to the noise transmitted through the engine structure. The noise, which sounds like a metallic ringing, different from the muted sound of normal combustion, is a consequence of the pressure waves generated by the abnormal combustion. The abnormal combustion suddenly releases a chemical energy that excites within the cylinder volume a local increase of pressure followed by oscillations. It is also generally accepted that spontaneous ignition begins at one or more exothermic centers in the end-gas region, as confirmed by optical probe measurements [11, 12]. Optical measurement methods are fulfilled requirements because the spontaneous nature of a self-ignition event and the high propagation velocity must be accounted for with adequate signal recording techniques, which detect in-cylinder gas signals and provide location of the knock signal origin [13–15].

Figure 2.3 shows typical knock conditions obtained in SI. The flame front profile has a concavity with a radius of curvature located in the opposite direction of the main propagation. In particular, SOI-307 (cycle n°57) flame outline presents more than one concavity, with a center of curvature at the bottom left or right of the chamber (intake valves). Instead, early split injection case (SOI-307-200) shows one wider concavity clearly visible in the exhaust valves, i.e., in the upper part of the frame. One large concavity and two smaller ones characterize the SOI-307-100 flame profile, due to the different flame speed previously described. Explosions causing knock often appear in close connection with a negative curvature in the flame front propagation profile. The presence of this negative curvature is associated with the slowdown of main flame front propagation and to low-temperature chemical kinetics in the zone not yet reached by normal combustion. Indeed, it was found out that the location of self-ignitions is strongly related to the curvature of the propagating flame front. This process can be deeply analyzed by evaluating the spatial evolution of the center of mass weighted on the luminous intensity (centroid of luminosity). Figure 2.4 illustrates the movement of the centroid from the spark plug position for one of the three analyzed conditions, SOI-307 cycle n°57. The centroid of luminosity starts moving slowly, and it remains located near the spark plug position. A slow deviation can be observed toward the exhaust or intake valves for all the test cases, reflecting the previously described prevalent flame front propagation directions. This shift can be partially masked by the carbonaceous deposits on the window.

Fig. 2.3 Comparison between flame outline before explosion and knock onset location in the end-gas region, for three selected test cases

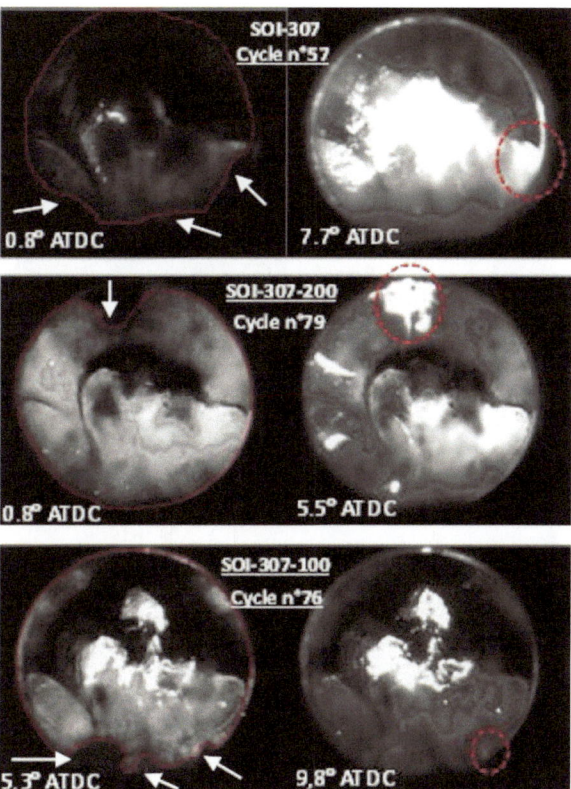

The spectroscopic measurements have allowed to characterize the flame evolution with high spatial and temporal resolution and to detect other species present during self-ignition, knock phase, and involved in pollutants formation. Several species involved in the first exothermic reactions due to the self-ignition of the air–fuel mixture can be detected by means of spectroscopic and ultraviolet (UV) [16]. Simultaneous use of spectral emissivity and absorption measurements in UV-visible range have shown to be a powerful tool comparing with high-speed imaging techniques. In fact, the ultraviolet spectrum is important because of the presence in this region of the well-known "water bands" due to OH radical that contributes to oxidation of fuel, knock marker [17], formation of NO_x [17], and oxidation of soot [18]. CH and C_2 bands provide a more precise signature of the reaction zone. CH can be used as an indicator of temperature in flame studies and C_2 as a CH precursor or precursor of soot [19]. Finally, CH/OH ratio intensity can be employed as a measure of the local equivalence ratio [20, 21], whereas CN radical emission can be considered as the marker of spark discharge in the combustion chamber [22, 23]. Moreover, carbonaceous material and large aromatic molecules exhibit absorption bands between 200 and 350 nm [9, 16, 19, 24, 25].

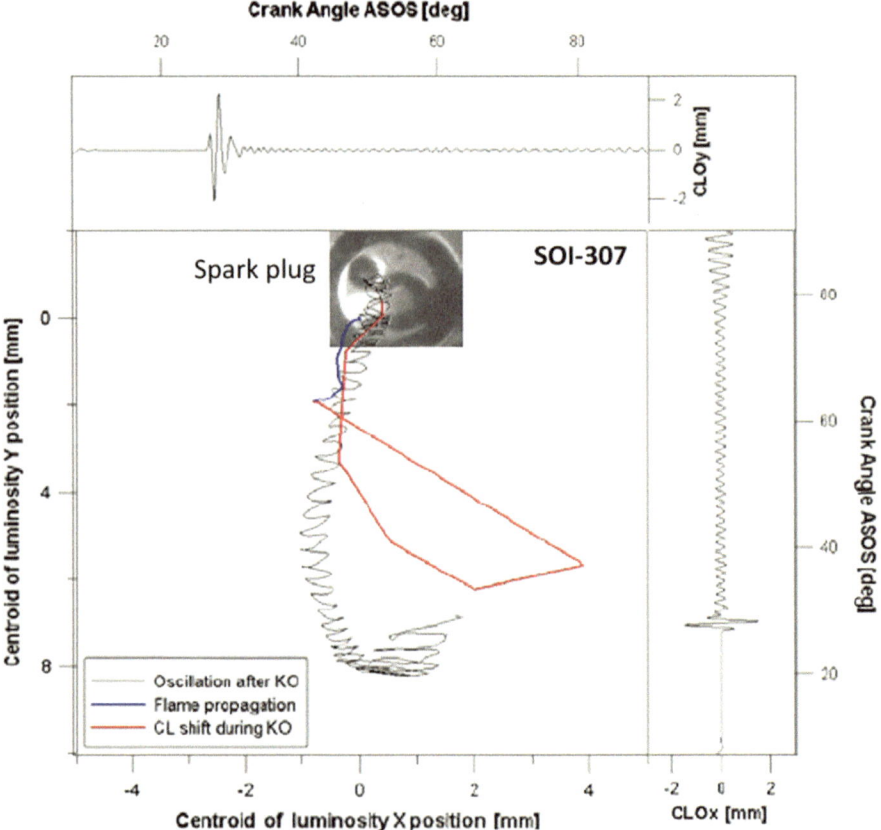

Fig. 2.4 Movement of centroid of luminosity evaluated for the test case SOI-307

2.4   Pollutants Formation

The luminosity/chemiluminescence imaging and the simultaneous planar imaging of laser-induced incandescence (LII) and elastic scattering have contributed to detect the onset of soot during the range of time in which a solid particle is formed from fuel molecules. The experimental difficulties involved in the soot formation and oxidation processes in the environment of the cylinder are enormous due to the high temperatures and pressures that extremely condition the reactive intermediate species occurring. Nevertheless, the light extinction, Rayleigh and Mie scattering, laser-induced incandescence (LII), and laser-induced fluorescence (LIF) allowed to follow the soot formation and oxidation process in terms of soot particle size and number density, as well as to have information on the flame temperature and the concentration of some typical species produced during the combustion process. As an example, NO concentration and its temporal evolution can be obtained by both absorption and LIF measurements, which involves closely linked advantages and disadvantages.

References

1. M. Astarita, F.E. Corcione, A. De Maio, B.M. Vaglieco, G. Valentino, An interpretation of high swirl diesel combustion based on optical diagnostics and 3D numerical calculations, in *Proceedings 14th International Symposium on Diagnostics and Modeling of Combustion in Internal Combustion Engines—COMODIA*, Kyoto (1998), pp. 149–154
2. H. Zhao, N. Ladommatos, *Engine Combustion Instrumentation and Diagnostics* (SAE International, Warrendale, PA, 2001)
3. E. Winklhofer, B. Ahmadi-Befrui, B. Wiesler, G. Cresnoverh, The influence of injection rate shaping on diesel fuel sprays—an experimental study. Proc. IMechE **206**, 173–183 (1992)
4. E. Mancaruso, L. Sequino, B.M. Vaglieco, Analysis of spray injection in a light duty CR diesel engine supported by non-conventional measurements. Fuel **158**, 512–522 (2015)
5. F. Schulz, J. Schmidt, A. Kufferath, W. Samenfink, Gasoline wall films and spray/wall interaction analyzed by infrared thermography SAE. Int. J. Engines **7**(3), 1165–1177 (2014)
6. M. Suzuki, K. Nishida, H. Hiroyasu, Simultaneous concentration measurement of vapor and liquid in an evaporating diesel spray. SAE Paper 930863 (1993)
7. H. Hiroyasu, M. Arai, M. Tabati, Empirical equations for the Sauter mean diameter of a diesel spray. SAE paper 890464 (1989)
8. M. Astarita, F.E. Corcione, B.M. Vaglieco, G. Valentino, Fuel composition effects on air-fuel mixing and self-ignition in a divided chamber diesel system by optical diagnostics. *Process in Diesel Engine Combustion*. SAE paper 1999-01-0510, vol. 1444 (1999)
9. E. Mancaruso, S.S. Merola, B.M. Vaglieco, Multi-injection combustion process in a transparent DI common rail diesel engine. Int. J. Engine Res. **9**(6), 483–498 (2008)
10. F. Catapano, P. Sementa, B.M. Vaglieco, Optical characterization of bio-ethanol injection and combustion in a small DISI engine for two wheels vehicles. Fuel **106**, 651–666 (2013)
11. G. Konig, C.G.W. Sheppard, End gas autoignition and knock in a spark ignition engine. SAE paper 902135 (1990)
12. U. Spiecher, L. Speigel, B. Reggelin, Investigation into applicability of an optical fiber sensor for knock detection and knock control system . SAE paper 922370 (1992)
13. H. Philipp, A. Plimon, G. Fernitz, A. Hirsch, G.K. Fraidl, E. Winklhofer, A tomographic camera system for combustion diagnostics in Si-engines. SAE paper 950681 (1995)
14. H. Philipp, A. Hirsch, M. Baumgartner, G. Fernitz, Ch. Beidl, W. Piock, E. Winklhofer, Localisation of knock events in direct injection gasoline engines. SAE paper 2001-01-1199 (2001)
15. F. Wytrykus, R. Dusterwald, Improving combustion process by using a high speed UV sensitive camera. SAE paper 2001–010917 (2001)
16. E. Mancaruso, B.M. Vaglieco, Spectroscopic measurements of premixed combustion in diesel engine. Fuel **90**(2), 511–520 (2011)
17. J.B. Heywood, Internal combustion engine fundamentals, vol. 930 (McGraw-Hill, New York, 1988)
18. G P. Merker, C. Schwarz, R. Teichmann, Combustion engines development (Springer, Berlin, 2012)

19. A.G. Gaydon, *The Spectroscopy of Flames* (Chapman and Hall, London, 1957)
20. F. Catapano, S. Di Iorio, P. Sementa, B.M. Vaglieco, Spectroscopic techniques for the evaluation of the in-cylinder air fuel ratio in a small optical Si engine fueled with methane and hydrogen/methane blends, in *2014 Fotonica AEIT Italian Conference on Photonics Technologies* (2014), pp. 1–4
21. S. Di Iorio, P. Sementa, B.M. Vaglieco, Experimental investigation on the combustion process in a spark ignition optically accessible engine fueled with methane/hydrogen blends. Int. J. Hydrogen Energy **39**(18), 9809–9823 (2014)
22. T.D. Fansler, M.C. Drake, B. Stojkovic, M.E Rosalik, Local fuel concentration, ignition and combustion in stratified charge spark direct injection engine: spectroscopic, imaging and pressure-based measurements. Int. J. Engine Res. **4**(2), 61–86 (2003)
23. S.S. Merola, S. Di Iorio, A. Irimescu, P. Sementa, B.M. Vaglieco, Spectroscopic characterization of energy transfer and thermal conditions of the flame kernel in a spark ignition engine fueled with methane and hydrogen. Int. J. Hydrogen Energy **42**(18), 13276–13288 (2017)
24. E. Mancaruso, L. Sequino, B.M. Vaglieco, C. Ciaravino, A. Vassallo, Spray formation and combustion analysis in an optical single cylinder engine operating with fresh and aged biodiesel. SAE paper 2011-01-1381, SAE Int. J. Engines 4(Issue 1), 1963–1977 (2011)
25. C.N.R. Rao, *Ultra-Violet and Visible Spectroscopy* (Butterworths & Co., London, 1977)
26. H. Hiroyasu, M. Arai, Structure of fuel sprays in diesel engines. Trans SAE. vol. 99, Sect. 3, pp. 1050–1061 (1990)
27. https://www.dantecdynamics.com/index
28. E. Mancaruso, B.M. Vaglieco, Spectroscopic analysis of the phases of premixed combustion in a compression ignition engine fuelled with diesel and ethanol. Appl. Energy **143**, 164–175 (2015)

Chapter 3
Modeling of Particle Size Distribution at the Exhaust of Internal Combustion Engines

Nowadays, the interest in the effect of exhaust emissions from road vehicles on public health is stronger than ever. Great attention is paid to particulate matter (PM) both for its impact on the environment and for the adverse effect on human health. The internal combustion engines (ICEs), both spark ignition (SI), and compression ignition (CI) are the main sources of PM emissions in the urban area. Particles are usually classified according to their diameter in coarse particles, diameter larger than 10 μm (PM10) and fine particles, diameter smaller than 2.5 μm (PM2.5). Further distinction is made for PM2.5, particles smaller than 100 nm are called ultrafine particles and those smaller than 50 nm are called nanoparticles. The chemical nature of the particles as well as the number and size depends on the engine type. Diesel engine particles consist mainly of agglomerated carbonaceous primary particles on which volatile organic material is adsorbed. The gasoline particles, instead, are mainly composed of organic fraction. Both CI and SI engines emit mainly particles in the ultrafine size range. Anyway, the particles' emissions from gasoline direct injection (GDI) engines are higher than that for port fuel injection (PFI) engines and Diesel engines equipped with a Diesel particulate filter (DPF).

The severe adverse effects on human health of fine and ultrafine particles emitted from internal combustion were well described in the literature [1–3]. Recent studies evidenced the strict relation between the particle size and the impact on heart and brain [4]. Smaller particles can, in fact, penetrate more easily the cell membranes than large particles [5]. Considering the negligible weight of the fine particle, a particle number (PN) emission limit is enforced in addition to the PM mass emission limits for particles larger than 23 nm at the Euro 6 (2014) for all categories of light-duty (LD) DI vehicles.

Great efforts are paid to reduce the particle emissions. Several solutions are under study, regarding the optimization of the combustion and the use of biofuel to reduce particle formation as well as the improvement of after-treatment devices for the reduction of emissions at the exhaust. In any case, availability of real-time information on the characteristics of particulate emissions, such as particle number and size, would enable the development of advanced closed-loop control

© The Author(s) 2018
F. Taglialatela Scafati et al., *Nonlinear Systems and Circuits in Internal Combustion Engines*, SpringerBriefs in Nonlinear Circuits,
https://doi.org/10.1007/978-3-319-67140-6_3

architectures aimed at reducing the impact of toxic particle emissions. However, modeling of the mechanisms of particle formation is not easy due to the complex and highly nonlinear processes involved.

The present chapter will firstly describe some of these mechanisms and, then, it will introduce a soft computing model, developed by the authors, devoted to the real-time prediction of particle size distribution at the exhaust of internal combustion engines on the basis of some specific inputs, such as engine speed, engine load, and amount of exhaust recirculated gases.

3.1 Particulate Matter Emissions in Engines

Particulate matter (PM) consists of tiny solid particles and liquid droplets that in urban area are mainly due to vehicles' emissions. Particles differ in size, composition, solubility, therefore also in their toxic properties. The composition of the particles depends on the fuel and on the combustion process. PM is usually composed of three basic fractions:

- Solids: dry carbon particles, commonly known as soot;
- Soluble organic fraction (SOF): heavy hydrocarbons adsorbed and condensed on the carbon particles;
- SO4: sulfate fraction, hydrated sulfuric acid.

A small amount of inorganic ash due to metal compounds in the fuel (if metallic additives are present) and lubrication can also be present.

Soot is formed from unburned fuel that nucleates from vapor to solid phase in fuel-rich regions at elevated temperatures. Hydrocarbons or other available molecules due to un/partially burned fuel/lubricant oil may condense on or be absorbed by soot depending on the surrounding conditions.

The evolution of liquid- or vapor-phase hydrocarbons to solid particles and possibly back to gas-phase products involves many chemical and physical processes that can be grouped into two steps: PM formation and PM oxidation. PM formation takes place via five different processes: pyrolysis, nucleation, coalescence, surface growth, and agglomeration. The nucleation and surface growth involve vapor concentrations and gas/particle conversion processes [6]. The agglomeration, instead, regards a solid phase. The step of particle formation and the scheme of PM formation are reported in Fig. 3.1 [6]. In practical combustion systems, the sequence of processes may vary between the two extremes.

A detailed scheme of soot formation and oxidation is depicted in Fig. 3.2.

Pyrolysis is the process whereby organic compounds, such as fuels, in the presence of high temperature alter their molecular structure without significant oxidation even though oxygen species may be present. All fuels undergo pyrolysis and produce essentially the same species: unsaturated hydrocarbons, polyacetylenes, polycyclic aromatic hydrocarbons (PAH), and especially acetylene. Fuel

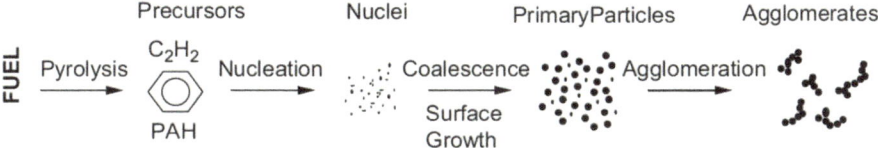

Fig. 3.1 Schematic diagram of the steps in the soot formation processes from gas to solid agglomerates particles

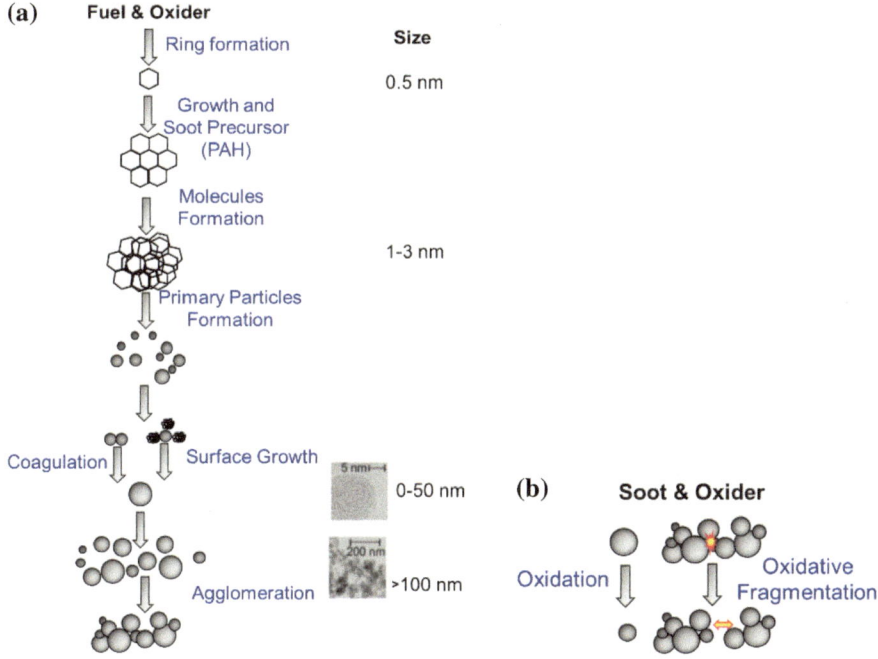

Fig. 3.2 Formation (**a**) and oxidation (**b**) mechanisms

pyrolysis results in the production of some species that are precursors or building blocks for soot. Soot precursor formation is a competition between the rate of pure fuel pyrolysis and the rate of fuel and precursor oxidation by the hydroxyl radical, OH. Both pyrolysis and oxidation rates increase with temperature, with the latter being faster. This explains why in premixed flames, where some amount of oxygen is present, soot is less than in respect to diffusion flames where no oxygen is present in the pyrolysis region.

Nucleation or soot particle inception is a homogeneous process occurring in the gas phase. The formation of particles from gas-phase reactants is controlled by chemical aspects. The particle inception process consists of radical additions of small, probably aliphatic, hydrocarbons to larger aromatic molecules [7]. Particle inception temperatures vary from 1300 to 1600 K. The nuclei of particles do not contribute significantly to the total soot mass, but have a significant influence on the mass added later, because they provide sites for surface growth.

The greater part of soot is formed by surface growth rather than soot inception. Surface growth is a process of adding mass to the surface of a nucleated soot particle. There is no clear distinction between the end of nucleation and the beginning of surface growth, and in reality, the two processes are concurrent. During surface growth, the hot reactive surface of the soot particles readily accepts gas-phase hydrocarbons. This leads to an increase in soot mass, while the number of particles remains constant. Surface growth continues as the particles move away from the primary reaction zone into cooler and less reactive regions, even where hydrocarbon concentrations are below the soot inception limit [8]. The problem is that surface growth is not a gas-phase reaction of small molecules, but a hetero-geneous process, where adsorption and desorption processes at the surface have to be considered as well. The majority of the soot mass is added during surface growth, and thus, the residence time of the surface growth process has a large influence on the total soot mass or soot volume fraction. Surface growth rates are higher for small particles than for larger particles because small particles have more reactive radical sites [9]. Coalescence and agglomeration are both processes by which particles combine.

Coalescence (sometimes called coagulation) occurs when particles collide and coalesce, thereby decreasing the number of particles and holding the combined mass of the two soot particles constant. During coalescence, two roughly spheri-cally shaped particles combine to form a single spherically shaped particle. Coagulation speed depends on particle diameter. Coagulation is more efficient between particles of different size since the speed of coagulation is a product of size and diffusion coefficient: big particles provide a large absorbing surface, and the smaller particles have a rapid diffusion [10]. Coagulation takes place only for relatively small particles (up to a diameter of 10 nm) which are characterized by high rates of growth in low-pressure premixed systems [11].

Agglomeration occurs when individual or primary particles stick together to form large groups of primary particles. The primary particles maintain their shape. Typically, the combined soot particles form chain-like structures, but in some cases, clumping of particles can be observed. Soot agglomeration takes place in the late phase of soot formation when, due to lack of surface growth, coagulation is no longer possible [12]. As a result, open-structured aggregates containing from 10 to 100 primary particles and characterized by a lognormal size distribution are formed [13].

The process of soot particle oxidation is parallel to the surface growth being itself a surface reaction. Soot particle oxidation occurs when the temperature is higher than 1300 K [14]. Once carbon has been partially oxidized to CO, the carbon will no longer evolve into a soot particle even if entering a fuel-rich zone. Oxidation can take place at any time during the soot formation process from pyrolysis through agglomeration. Potential soot oxidants are O, O_2, OH, and CO_2. The major oxi-dation process occurs at the very beginning of soot particle growth, which is the soot particle nucleation period, where a rapidly decreasing concentration of O_2 in fuel-rich environments is observed. The most active oxidation species depends on the process and state of the mixture. Oxidation of small particles is considered a two-stage process. First a chemical attachment of oxygen to the surface (absorption) occurs, then a desorption of the oxygen with the attached fuel component from the surface as a product takes place [15]. Under lean conditions, soot is oxidized by both OH and O_2 [16]. OH is most likely to dominate soot oxidation under fuel-rich and stoichiometric conditions [17] as the soot decreases with an increase in OH concentration. CO_2 addition has an indirect effect on soot oxidation. CO_2 promotes hydroxyl concentration that in return increases the oxidation of soot precursors.

The PM are not spherical particles, but they are aggregates and contain void spaces. If their material density is unknown, in order to determine their size it is necessary to know one or more parameters. For this reason, often it is most convenient refer to equivalent diameters, defined as the diameter of a sphere, which with a given instrument would yield the same size measurement as the particle under consideration [18].

The most commonly used equivalent diameters are as follows:

- Volume equivalent diameter (dve): also known as envelope equivalent diameter, is defined as the diameter of a spherical particle of the same volume as the particle under consideration. For an irregular particle, dve is the diameter that the particle would have if it were melted to form a droplet while preserving any internal void spaces. dve is equal to the geometric diameter dp for spherical particles [19].
- Mass equivalent diameter (dme): is similar in concept to dve with the difference that dme does not include internal voids. Therefore, for a particle with no internal voids dve = dme. If the particle contains internal voids, dve > dme.
- Electrical mobility diameter (dm): is the diameter of a sphere with the same migration velocity in a constant electric field as the particle of interest. Instruments such as the DMS and the SMPS measure dm. This measurement is obtained via a force balance between the electrical force of a constant electric field on the net charges on the particle and the drag force experienced by the particle.
- Aerodynamic diameter (da): is the diameter of a spherical particle having a density of 1 g/cm^3 that has the same terminal settling velocity in the gas as the particle of interest. Terminal settling velocity is a measure of the aerodynamic properties of the particle, and the aerodynamic diameter depends on the flow regime. Instruments such as the ELPI measure da. The aerodynamic diameter has been developed in order to provide a simple means of categorizing the sizes of particles having different shapes and densities with a single dimension.
- Stokes diameter (ds): is the diameter of a sphere which has the same density and setting velocity as the particle of interest [20].

For irregular particles of standard density, dm < dve < da.
According to their diameter, particles are usually classified in [21]:

- Nano: particles smaller than 50 nm;
- Ultrafine: particles with an aerodynamic diameter lower than 0.1 μm;
- Fine (PM2.5): particles with an aerodynamic diameter in the size range from 0.1 up to 2.5 μm;
- Coarse (PM10): particles with an aerodynamic diameter in the size range from 2.5 up to 10 μm;
- Supercoarse: particles with an aerodynamic diameter higher than 10 μm.

The particles usually form separated modes in the number size distribution. They can be classified as follows:

- Nuclei mode: particles smaller than 50 nm, usually formed from a volatile precursor during exhaust dilution and cooling processes;
- Accumulation mode: particles from 50 nm up to 1000 nm usually consisting of carbonaceous agglomerates and adsorbed material;
- Coarse mode: particles larger than 1000 nm consisting of re-entrained accumulation mode particles and crankcase fumes.

However, the last generation engines usually emit only particles in nucleation and accumulation mode. Figure 3.3 shows a typical size distribution of atmospheric particulate matter of a Diesel engine, where no particles in coarse mode are emitted.

Highly irregular particle populations, such as PM, will show significant differences in the size distributions measured simultaneously by mobility and aerodynamic techniques. These are not real discrepancies; instead, they merely capture the different dependence of both equivalent diameters on the fundamental particle properties. This phenomenon has been observed in several investigations [22, 23, 24].

PM formation process is strictly affected by the combustion process. Soot concentration, in fact, depends on C/O ratio, therefore on the equivalence ratio of

Fig. 3.3 Typical particle size distribution

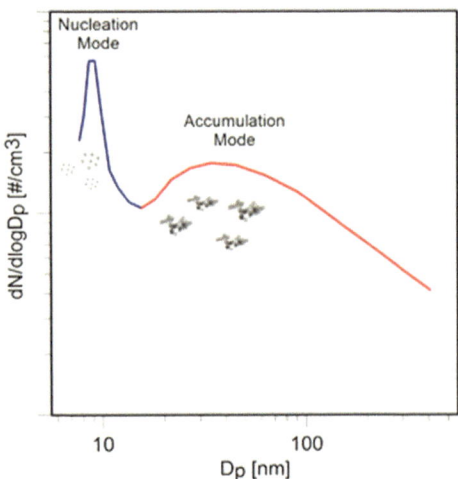

the mixture, apart from temperature and pressure. For C/O ratio bigger than a critical value, soot formation occurs in the temperature range from 1400 to 2200 K, with a maximum at around 1700 K.

Diesel particulate matter is mainly composed of carbonaceous particles generated in the engine cylinder during the combustion. The primary particles are carbonaceous spherical in shape on which are adsorbed unburned compounds that agglomerate to form long chain-like structures. In details, small nuclei of hydrocarbon with low H/C ratio are formed where combustion is locally rich. Then, these nuclei particles aggregate and adsorb smaller particles to their surface. Diesel soot formation mechanism is described in [24].

According to Dec theory, soot is formed throughout the fuel jet cross section, but the major part of soot is oxidized inside the cylinder. The actual composition of Diesel PM depends on the engine type and load and speed conditions. "Wet" particles can contain up to 60% of the hydrocarbon fraction (SOF), while "dry" particles comprise mostly of dry carbon. The amount of sulfates is directly related to the Sulfur contents of the fuel [25]. Primary soot particle size appears to vary depending on operating condition, injector type, and injector conditions but most primary particles' sizes reported range from 10 to 70 nm. After combustion ends, particles agglomerate and are seen to be chain-like.

The mechanism of particle formation in Gasoline Port Fuel Injection Spark Ignition (GPFI SI) engines is very different from Diesel engines and not yet well understood. In a PFI SI engine, fuel and air are premixed in the intake duct before combustion. Particle formation in this case results from local inhomogeneous conditions, fuel pooling, big droplets, cracks, crevices, etc. Analyses of particulate emissions from gasoline PFI engines reveal that the bulk of the mass is ash with the second largest fraction of unburned lubricating oil. Carbon emissions are found to be significant only at high speed and load with mixture enrichment. PM emission from PFI SI engines is negligible with respect to the Diesel ones and, for this reason, they are not considered in the actual emission standard limits. It was observed that gasoline particles' emissions depend on engine speed and load [26, 27]. Nevertheless, there are few detailed analysis on the effect of the state variables, such as rpm and load, and control parameters, such as AFR and spark timing, on particles number and size.

For gasoline direct injection (GDI) engines, instead, the environmental condition prompt to particle formation [28]. Differently, from GPFI SI engines, less time is available for the charge to become a homogeneous mixture and even for all the fuel to evaporate. In this case, in fact, the time available for mixture preparation and fuel evaporation is limited to the window between injection and ignition. The consequent charge heterogeneity and fuel droplet presence are two mechanisms causing large PM emissions in GDI engines. Another aspect to consider is the piston and wall wetting due to the fuel impingement on the piston crown creating a fuel film. This film can burn diffusively and as a pool fire. Both cases are particularly effective at forming primary carbon particles that later aggregate.

Anyway, whatever the CI and SI configuration, the engines emit a large number of ultrafine particles. Health studies [2] have shown that ultrafine particles have a greater adverse effect on men's health than larger particles of the same composition since they are capable of penetrating deeper into the respiratory tract. Furthermore, particle dynamics and lifetime in the atmosphere can influence visibility, radiative balance, climate change, with the extent dependent on the size distribution and nature of the aerosols [29].

The larger thoracic coarse particles are primarily deposited in the nose and in larger airways due to impaction because they cannot follow the air stream at bifurcations; these particles are filtered out of the air stream and cannot penetrate down to the deep lung. Smaller particles can pass through the large airways and are deposited in the lung due to sedimentation (settlement of particles in resting air due to gravity). These fine particles may preferentially affect the cardiovascular system. For this reason, particle regulation on the number of particles larger than 23 nm was introduced from Euro 6 for direct injection light-duty cars [30].

3.2 Real-Time Prediction of Particle Sizing at the Exhaust of a Diesel Engine

The present paragraph summarizes the results of studies carried out by the authors on the prediction of characteristics of exhaust particles emitted by Diesel engines. In particular, it will be described a neural network model that provides real-time information about particle number and size on the basis of engine parameters such as angular speed, load, and EGR level.

The experimental activity was carried out on a four-stroke Diesel engine, with four inline cylinders, 16 valves, a displacement of 1.9 l, and a compression ratio of 17.5:1. The engine was equipped with a common rail injection system, which allowed two injections (Pilot—Main) for each cycle (Unijet). The engine was also provided with an exhaust gas recirculation (EGR) system, whose basic element is an electro-valve controlled by the electronic control unit. The electro-valve receives an electrical signal (duty-cycle) and orders a pressure reduction to the mechanical valve (EGR valve) placed between the outlet and the inlet pipes. When the depression is applied, the EGR valve allows the addition of part of exhaust gases to the airflow. The whole process is controlled by the electronic control unit using characteristic look-up tables stored in its memory. The actual EGR percentage, according to the conventional definition [31], can be obtained measuring the aspired air ($\dot{m}_{air}$) with and without EGR and using the following equation:

$$\text{EGR}\% = 1 - \frac{\dot{m}_{\text{air wEGR}}}{\dot{m}_{\text{air w/oEGR}}} \cdot \frac{p_{\text{w/oEGR}}}{p_{\text{wEGR}}} \cdot \frac{T_{\text{wEGR}}}{T_{\text{w/oEGR}}}, \tag{3.1}$$

in order to get the exact EGR percentage values, corrections of pressure and temperature are necessary. The presence of exhaust gases in the inlet manifold does not imply the same reduction of the aspired fresh air. This is due to the contribution of exhaust gases thermal energy, and it leads to an increase of the aspired fluid temperature and therefore to its expansion.

An opacimeter was used to evaluate particulate mass concentration as a function of time. Opacimeter is a partial-flow system that measures the visible light attenuation (550 nm) from the exhaust gases. From empirical relations, it is possible to convert the opacity percentage in particulate mass concentration [32].

Particle size distribution at the Diesel engine exhaust was evaluated by means of an electrical low-pressure impactor (ELPI). It measures in real time the particle aerodynamic diameter in the range 7 nm–10 μm [33]. The sample first passes through a unipolar positive polarity charger, where the particles are electrically charged by ions produced in a corona discharge. After the charger, the particles pass in a low-pressure impactor where they are classified according to their aerodynamic diameter. The stages of the impactor are electrically insulated, and each stage is individually connected to an electrometer current amplifier. The charged particles collected in a specific impactor stage produce an electrical current, which is recorded by the respective electrometer channel. The current value of each channel is proportional to the number of particles collected, and thus to the particle concentration in the specific size range. The current values are converted to an aerodynamic size distribution using particle-size-dependent relations describing the properties of the charger and the impactor stages [34]. Fine particle sampler (FPS) was used to sample and dilute ratio the exhaust gas.

The modeling approach used for prediction of particulate emissions characteristics was based on the idea of developing a soft computing model acting as a "supervisor" of PM emissions from a Diesel engine. In particular, the model gives, for every engine operating condition, information about the characteristics of particulate emissions in terms of particle size distribution.

The soft computing model has as output the number concentration of particulate particles with fixed aerodynamic diameters, respectively, of 8, 28, 54, 91, 154, 261, and 381 nm, whereas the inputs are the engine speed, the engine load, and the actual EGR ratio (see Figure 3.4). The proposed model has been conceived taking into account the paradigm of a learning machine trained and tested on experimental data. As learning machine model, a feedforward MLP (Multilayer Perceptron) neural network with one hidden layer has been chosen.

To improve the generalization capability of the neural network model, an automated regularization procedure applied during the model training, known in the literature as Bayesian regularization, was considered, see [35, 36]. According to the Bayes's rule, the "best" model is defined as the model with the highest a posteriori probability of the correctness, this means:

Fig. 3.4 Block scheme of the
model for prediction of
particulate emissions
characteristics

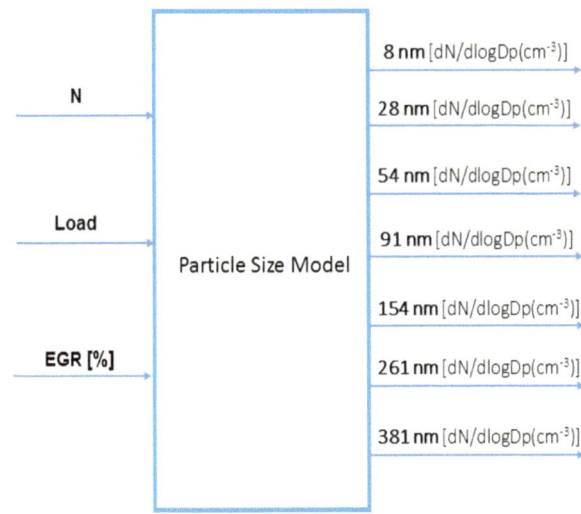

$$\min_{\theta}(-\log(P(D|H_\theta) - \log(P(H_\theta))), \tag{3.2}$$

where θ is the parameter of the model H_θ used to describe a system, whose observed event is indicated with D. In Eq. (3.2), $P(D|H_\theta)$ represents the probability of the event conditioned by the correctness of the model. In the context of a neural network model, the observed event of the system is:

$$D = \{(\bar{x}^i, \bar{y}_i)|i = 1, \ldots, n\}, \tag{3.3}$$

where $\bar{x}^i$ is the i-th input vector of the neural network model, and $\bar{y}_i$ is the related i-th output vector used as dataset for the model training. The neural network is described as a nonlinear map g_θ:

$$\bar{y}_i = g_\theta(\bar{x}^i), \tag{4}$$

which, in function of the weights and biases of the hidden layer neurons (all described by the vector θ), links the input–output data. In the Bayesian neural network approach, a normal Gaussian data distribution is supposed only if the model is correct. So, if θ is the correct parameter vector, the observed values $\bar{y}_i$ are normally distributed around $g_\theta(\bar{x}_i)$. After trivial calculations, it can be obtained that:

$$
\begin{aligned}
-\log(P(D|H_\theta)) &= -\log \prod_{i=1}^{n} e^{-\frac{(\bar{y}_i - g_\theta(\bar{x}^i))^2}{2r^2}} \\
&= \frac{1}{r^2} \sum_{i=1}^{n} \frac{1}{2} (\bar{y}_i - g_\theta(\bar{x}^i))^2,
\end{aligned} \tag{3.4}
$$

therefore, the function to be minimized (during the training phase of the model) is:

$$\min_{\theta} \left(\sum_{i=1}^{n} \frac{1}{2} \left(\bar{y}_i - g_{\theta}(\bar{x}^i) \right)^2 - r^2 \log(P(\theta)) \right), \tag{3.5}$$

with $-r^2 \log(P(\theta))$ that represents a kind of a priori information about the correct solution.

A dataset composed of 38 experimental points was used to train and test the neural network model. The whole dataset was then divided into training and testing sets, composed of 28 and 10 experimental points, respectively. For each particle aerodynamic diameter, training and testing datasets included engine speeds ranging from 1000 to 2000 rpm, engine loads from 2 to 5 bar, and EGR percentages from 0 to 56%. Training data were chosen in order to contain information spread evenly over the entire range of engine operative conditions. This in order to increases generalization capability of the neural network model. The number of neurons in the hidden layer was set equal to 5. Also, other neural network configurations were tested. In particular, maintaining a fixed number of neurons in the hidden layer, the performances of the model to changes in the ratio between training and testing dataset size (from 24/14 to 33/5) were evaluated. Moreover, the influence of the number of neurons in the hidden layer on the performances of the model was also investigated. The results of all these tests confirmed that the model configuration with 5 neurons in the hidden layer and a ratio between training and testing dataset equal to 28 to 10 provide the lower absolute mean square error in the prediction of particle concentration, for all the particle equivalent diameters considered.

Figure 3.5 shows, for different EGR ratios, the particle size distribution measured at the exhaust of the test engine at 1500 rpm and 30 Nm.

Fig. 3.5 Particle size distribution at different EGR ratio

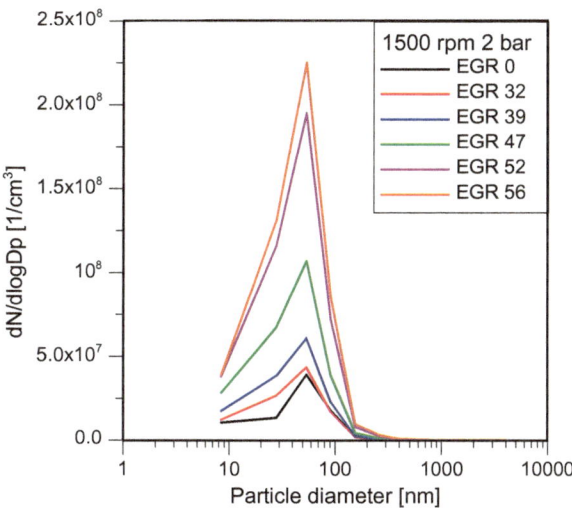

Table 3.1 Absolute and percent mean square error related to the estimated concentration of the particles having as equivalent diameters, respectively, 8, 28, 54, 91, 154, 261, and 381 nm

Particle diameters (nm)	Absolute mean square error (cm^{-3})	Percent mean square error $(\%)$
8	5.0E+06	6
28	1.02E+07	4.4
54	2.36E+07	3.1
91	1.30E+07	4.3
154	2.14E+06	3.8
261	5.11E+05	3.3
381	2.51E+05	6

The figure clearly indicates that EGR does not affect significantly the particle size distribution, which for all the cases shows a unimodal behavior. However, an increase in EGR produces an increase in the number concentration. The particle formation is prompt in all the size range; however, the formation of larger particles is enhanced because of the lower temperature, which boosts the growing processes, such as coagulation and agglomeration, and of the lower presence of oxygen, which improves particle formation and worsens the particle oxidation. This result became more evident at the increasing of the EGR.

Table 3.1 shows the comparison between the experimental and the estimated number concentration of the selected size particles for the testing points chosen to validate the model.

The mean square error is within the range 3–6% of the maximum experimental data. This shows a satisfactory generalization capability of the neural network model; hence, the soft computing approach followed in this works seems to be able to guarantee similar fair performance also in other un-experimented engine conditions. Only for the lower and larger diameter, the error was higher than 5%, this can be due to the fact that these are the upper and lower limits of the instruments, so they measure can be slightly biased.

References

1. K. Donaldson, X.Y. Li, W. MacNee, Ultrafine (nanometer) particle mediated lung injury. J. Aerosol Sci. **29**, 553–560 (1998)
2. G. Oberdörster, M.J. Utell, Ultrafine particles in the urban air: to the respiratory track and beyond. Environ. Health Perspect. **110**, A440–A441 (2002)
3. C.A. Pope, D.W. Dockery, Health effects of fine particulate air pollution: lines that connect. J. Air Waste Manag. Assoc. **56**, 709–742 (2006)
4. I.M. Kennedy, The health effects of combustion-generated aerosols. Proc. Combust. Inst. **31**, 2757–2770 (2007)

5. K.A. Bérubé, T.P. Jones, B.J. Williamson, C. Winters, A.J. Morgan, R.J. Richards, Physicochemical characterisation of Diesel exhaust particles: factors for assessing biological activity. Atmos. Environ. **33**, 1599–1614 (1999)
6. A. Seaton, K. Donaldson, Nanoscience, nanotoxicology, and the need to think small. The Lancet; London **365** (9463), 923–4 (2005)
7. N. Ladommatos, R. Balian, R. Horrocks, L. Cooper, The effect of exhaust gas recirculation on soot formation in a high-speed direct-injection Diesel engine. SAE Technical Paper 960841
8. D. Ivaldi, M.G. Lisbona, M. Tonetti, An improved EGR system concept for Diesel engines towards fuel neutral emissions. Int. J. Veh. Des. **41**(1/2/3/4), 307–325 (2006)
9. D.R. Tree, K.I. Svensson, Soot processes in compression ignition engines. Prog. Energy Combust. Sci. **33**, 272–309 (2007)
10. W. Bartok, A.F. Sarofim, *Chemistry of gaseous pollutant formation and destruction, fossil fuel combustion* (Wiley, New York, 1991)
11. B.S. Haynes, H.G. Wagner, Soot formation. Prog Energy Combust Sci. **7**, 229–73 (1981)
12. M. Ketzela, R. Berkowicz, Modelling the fate of ultrafine particles from exhaust pipe to rural background: an analysis of time scales for dilution. Coagul. Deposition Atmos. Environ. **38**, 2639–2652 (2004)
13. B. Howard, B.L. Wersborg, G.C. Williams, Coagulation of carbon particles in premixed flames. Faraday Symp. Chem. Soc. **7**, 109–119 (1973)
14. J. Warnatz, U. Maas, R.W. Dibble, *Physical and chemical fundamentals, modelling and simulation, experiments, pollutant formation*, 4th edn. (Springer, Berlin, 2006)
15. O.I. Smith, Fundamentals of soot formation in flames with application to Diesel engine particulate emissions. Prog. Energy Combust. Sci. **7**, 275–291 (1981)
16. K.G. Neoh, J.B. Howard, A.F. Sarofim, Effect of oxidation on the physical structure of soot. Proc. Comb. Inst. **20**, 951–957 (1985)
17. P. DeCarlo, D.R. Worsnop, J.G. Slowik, P. Davidovits, J.L. Jimenez, Particle morphology and density characterization by combined mobility and aerodynamic diameter measurements. Part 1: theory. Aerosol Sci. Technol. **38**(12), 1185–1205 (2004)
18. W.C. Hinds, *Aerosol technology: properties, behavior, and measurement of airborne particles* (Wiley, New York, 1999)
19. J. Heyder, J. Gebhart, W. Stahlhofen, Water, diameters of airborne particles. Air Soil Pollut. 567–572 (2004)
20. ELPI User Manual, Dekati Ltd. Tampere, Finland (2006)
21. M. Matti Maricq, D.H. Podsiadlik, R.E. Chase, Size distributions of motor vehicle exhaust PM: a comparison between ELPI and SMPS measurements. Aerosol Sci. Technol. **33**, 239–260 (2000)
22. N. Collings, B.R. Graskow, Particles from internal combustion engines- what we need to know. Philos. Trans. R. Soc. Lond. A. **358**, 2611–2623 (2000)
23. C. Van Gulijk, J.C.M. Marijnissen, M. Makkee, J.A. Moulijn, A. Schmidt-Ott, Measuring Diesel soot with a scanning mobility particle sizer and an electrical low-pressure impactor: performance assessment with a model for fractal-like agglomerates. Aerosol Sci. **35**, 633–655 (2004)
24. J.E. Dec, A conceptual model of DI Diesel combustion based on laser-sheet imaging. SAE Paper 970873 (1997)
25. D.B. Kittelson, W.F. Watts, J.P. Johnson, Ultrafine and Nanoparticle Emissions: A New Challenge for Internal Combustion Engine Designers. International Conference "Engines of Sustainable Development", Naples, 2003
26. B.R. Graskow, D.B. Kittelson, M.R. Ahamadi, J.E. Morris, Exhaust particulate emissions from two port fuel injected spark ignition engines. SAE Paper n. 1999-01-1144 (1999)
27. T. Gupta, A. Kothari, D.K. Srivastava, A.K. Agarwa, Measurement of number and size distribution of particles emitted from a mid-sized transportation multipoint port fuel injection gasoline engine. Fuel **89**(9), 2230–2233 (2010)
28. F. Zhao, D.L. Harrington, M.-C.D. Lai, *Automotive Gasoline Direct-Injection Engines* (SAE, Warrendale, PA, 2002)

29. G. Oberdörster, E. Oberdörster, J. Oberdörster, Nanotoxicology: an emerging discipline evolving from studies of ultrafine particles. Environ. Health Perspect. **113**(7), 823-839 (2005)
30. Regulation (EC) No. 715/2007 of the European Parliament and of the Council on type approval of motor vehicles with respect to emissions from light passenger and commercial vehicles (Euro 5 and Euro 6) and on access to vehicle repair and maintenance information (2007)
31. S. Canale, R. Scotton, "Calcolo della percentuale di EGR – Motori – Sistemi Diesel ed Emissioni", Orbassano, January 1992
32. O. Mörsch, P. Sorsche (DaimlerChrysler AG), Investigation of Alternative Methods to Determine Particulate Mass Emissions, http://www.oica.net/htdocs/Main.htm
33. M. Maricq, N. Xu, E. Chase, Measuring particulate mass emissions with the electrical low pressure impactor. Aerosol Sci. Technol. **40**, 68–79 (2006)
34. H. Zhao, N. Ladommatos, Engine combustion instrumentation and diagnostics. SAE International, 2001
35. D.J.C. MacKay, Bayes interpolation. Neural Comput. **4**(3), 415–447 (1992)
36. F.D. Foresee, M.T. Hagan, Gauss-Newton approximation to Bayesian regularization, Proceedings of the 1997 International Joint Conference on Neural Networks, pp. 1930–1935 (1997)

Chapter 4
Diagnosis and Control of Engine Combustion Using Vibration Signals

In other parts of this book, the importance of non-intrusive diagnostic techniques to evaluate the combustion quality in internal combustion engines has been highlighted. Non-intrusive diagnostics offer several advantages: the sensors are generally placed externally the engine, therefore, no engine structural modifications are required. As a consequence, the sensors used for this kind of applications are not requested to resist very high temperature and pressure and they are relatively cheap. Among non-intrusive diagnostic techniques, those based on the analysis of the engine vibration seem to be the most promising and several studies may be found in the literature [1–12]. In particular, the strong connection between the characteristics of the combustion process and the so-called vibration signature of the engine induced some authors to perform a reconstruction of the cylinder pressure signal by using the information coming from single or multiple accelerometers placed on the engine block. The signal processing tools used for that purpose were de-convolution methods [1], spectrum analysis [2], cyclo-stationarity properties [5], and methods using neural network models [6].

The present chapter proposes a method for "advanced" combustion diagnosis and control without using in-cylinder pressure transducers. The method includes a vibration signal processing in order to separate the combustion phenomena from all the other noise signatures on the signal. The strict correlation between the filtered block vibration signal and some combustion parameters (such as MFB50 and combustion pressure peak location) has been also demonstrated. Finally, the chapter contains some possible combustion control strategies based on vibration signal analysis.

In order to test the proposed approach, several types of ICEs were employed. In particular, two single cylinder research engines (one compression ignition (CI) and another spark ignited (SI)), and two multi-cylinder engines, still compression and spark ignition, were used. Engines characteristics and injection system specifications are reported in Table 4.1. All engines had four valves per cylinder and were equipped with different injection systems, direct injection common rail for compression ignition engines, and indirect injection port fuel injection for spark

© The Author(s) 2018
F. Taglialatela Scafati et al., *Nonlinear Systems and Circuits in Internal Combustion Engines*, SpringerBriefs in Nonlinear Circuits,
https://doi.org/10.1007/978-3-319-67140-6_4

Table 4.1 Main features of the engines used for the experimental tests

	SCRE 1 CI	SCRE 2 SI	MCRE 3 CI	MCRE 4 SI
# of Cylinders	1	1	4	4
Bore (mm)	85	79	83	71
Stroke (mm)	92	81	90	79
Displacement (cm^3)	522	399	1956	1242
Compression ratio	16.5:1	10.0:1	16.5:1	10.1:1
Injection system	Common rail	Port fuel injection	Common rail	Port fuel injection

ignition. The injection systems were managed by ECUs that allowed to control the injection pressure, the injector energizing time, the injection timing, the number of injections per cycle, and the dwell time between consecutive injections.

All the engine vibration measurements were carried out by a low-cost, low-power, linear capacitive accelerometer working on a 6 kHz bandwidth. The accelerometer included a sensing element and an integrated circuit interface able to take the information from the sensing element and to provide an analog output signal. The sensor was set by means of glue in the upper surface of the elongated cylinder, close to the cylinder head, and under the exhaust manifold. This solution was found to be the best in terms of sensitivity and accessibility. The investigated engine operating conditions were at different engine speeds and brake mean effective pressure (BMEP).

In Fig. 4.1, the motored and fired in-cylinder pressure signals and the corre-spondent engine vibration signals are reported for three consecutive cycles of a Diesel engine at an engine angular speed of 1500 rpm. The accelerometer signal detected during the engine operation in motored condition provides important information about the background vibrations and it can be assumed as baseline. As can be noted at the bottom of Fig. 4.1, when the injection is activated and the combustion starts, the vibration signal characteristics vary with respect to the baseline. The difference between the vibration signal in motored and fired condi-tions occurs around the peak pressure and the combustion event is the main responsible of this variation.

Basically, engine block vibration is due to several sources such as valve opening and closing, piston slap, combustion process and several additive noises. Transient waves generated by these sources overlap each other and the main issue is to isolate the contribution of interest. In order to extract information about the combustion pressure from the vibration signal, the accelerometer signal has been analyzed by means of a time–frequency analysis.

In Fig. 4.2, the time–frequency study of the vibration signal is shown for both SCRE 1 CI engine (left) and SCRE 2 SI engine (right), respectively. The diagram has been obtained by using the MATLAB command "spectrogram": it produces time–frequency diagrams by making many STFTs (Short Time Fourier Transforms) of the same length. The amplitude of STFTs is then translated to a colorful graph. The spectrogram reported in Fig. 4.2 refers to a speed of 1500 rpm for CI engine

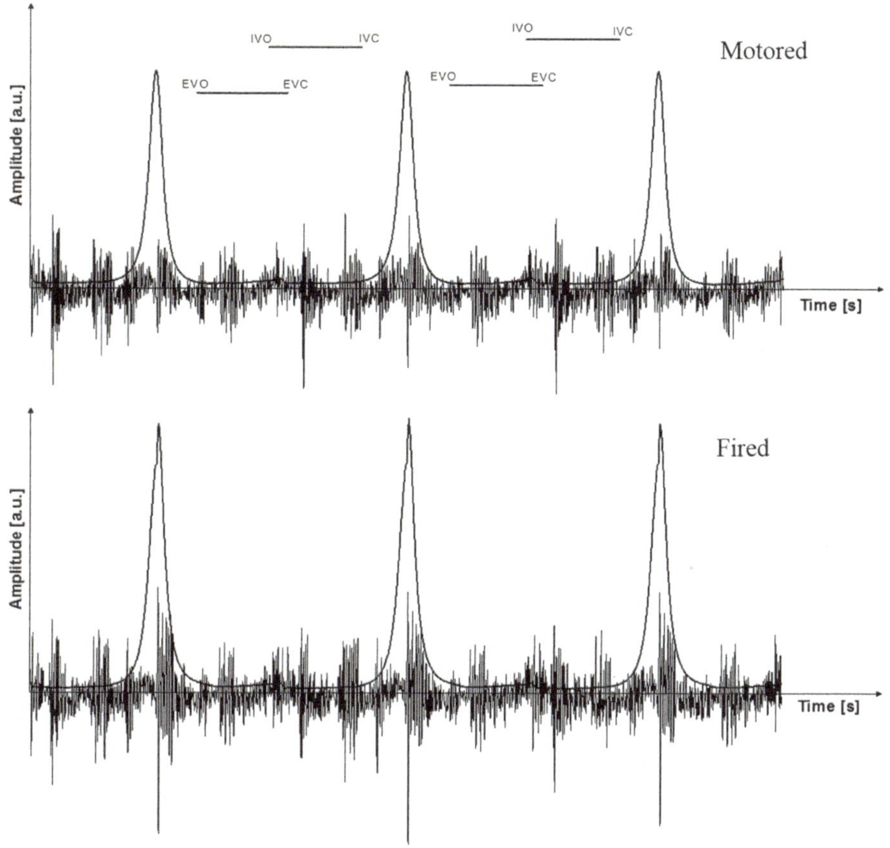

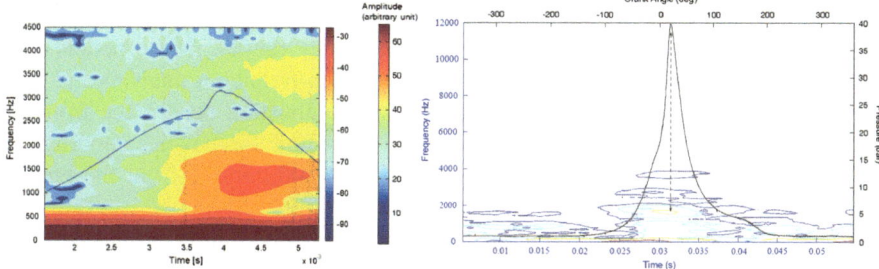

Fig. 4.1 Generic sampling of the in-cylinder pressure and accelerometer signals with engine in motored (*up*) and fired (*bottom*) conditions, respectively, at 1500 rpm

Fig. 4.2 Time–frequency spectrogram of accelerometer signals and averaged pressure cycles for SCRE1 engine (*left*) and SCRE 2 engine (*right*)

and 2000 rpm for SI engine. In order to help the identification of correlations between cylinder pressure cycle and vibration signal, also the average pressure cycle (averaged on 400 consecutive engine cycles), has been superimposed on the graphs.

From the analysis of the spectrogram, it can be deduced that the energy of combustion is prevailing in the frequency band 1000–1500 Hz for CI SCRE 1 engine. On the other hand, the frequencies below 4000 Hz well represented the whole combustion process for SCRE 2 engine. In particular, for this latter engine a dominant frequency band can be observed in the range 1500–1800 Hz. The high-amplitude components below 500 Hz for SCRE 1 and below 100 Hz for SCRE 2 are probably due to spurious sources and they cannot be associated to any combustion phenomena. Similar results were obtained with the time–frequency study performed at other engine speeds.

In order to select the frequency bandwidth of interest, it can be combined the time–frequency analysis with a coherence analysis. The value of the coherence function between the vibration signal and the cylinder pressure, in fact, gives frequencies where the information of combustion can be extracted from the vibration trace. The coherence function can be computed as a normalized measure that depends on the cross-power spectral density, $P_{p,v}(f)$, of cylinder pressure and block vibration signal, and on the spectral densities, $P_{p,p}(f)$ and $P_{v,v}(f)$, of each signal (see Eq. (1)).

$$C_{p,v}(f) = \frac{|P_{p,v}(f)|^2}{P_{p,p}(f).P_{v,v}(f)}. \tag{1}$$

$C_{p,v}(f)$, has values between 0 and 1, and higher is $C_{p,v}(f)$, higher is the correlation between cylinder pressure and block vibration signal.

Figure 4.3 shows the coherence between in-cylinder pressure and accelerometer signal for the SCR1 engine tested at 2000 rpm. From this figure, highest coherence values are exhibited in the frequency bandwidths 800–1200 Hz for SCRE 1. This indicates the strong correlation between the spectral components of in-cylinder pressure and accelerometer signals in these frequency ranges. These results also confirm the information obtained by means of the time–frequency analysis and they allow to define a band-pass filter to separate the combustion occurrence from all other noise sources on the vibration signal. The cutoff frequencies reported in Table 4.2 were chosen to filter the accelerometer signal coming from the engine block of the different engines used for experimental tests.

In Fig. 4.4, the angular position of maximum amplitude of filtered accelerometer signals (LMA) is plotted versus the angular location of the pressure peak (LPP) for all the engine operating conditions of the two multi-cylinder (MCR) engines tested. The graph shows a good linear-like correlation. In Table 4.3, averaged LMA values and the correspondent averaged LPP values are reported for all selected engine operating conditions. The standard error related to the estimation of the LPP by

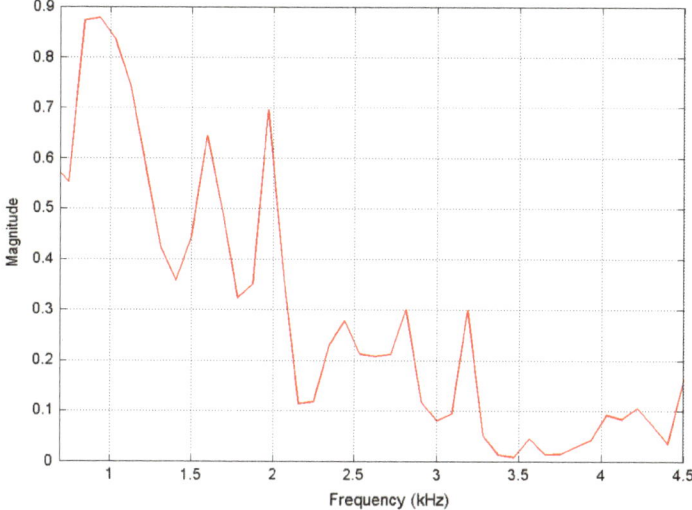

Fig. 4.3 Coherence analysis for SCRE 1 engine at 2000 rpm

Table 4.2 Frequency-ranges adopted to filter vibration signals

	1500 rpm (Hz)	2000 rpm (Hz)
SCRE 1 CI	1000–1500	800–1200
SCRE 2 SI	1500–1800	1500–1800
MCRE 3 CI	1500–3000	1500–3000
MCRE 4 SI	500–800	500–800

Fig. 4.4 Pressure peak and vibration signal maximum location for MCRE 3 and MCRE 4

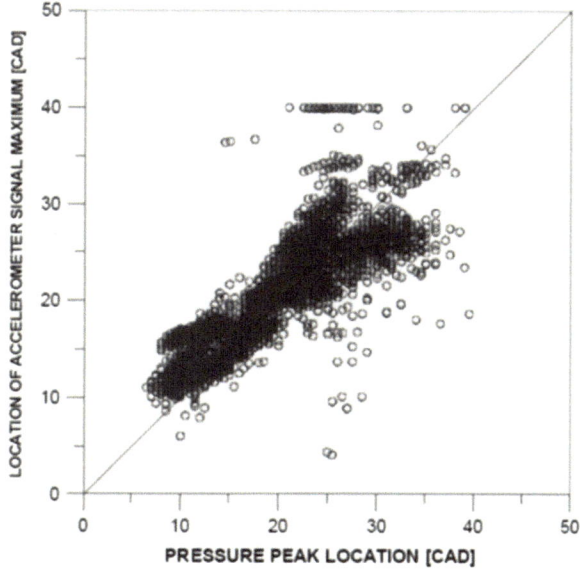

Table 4.3 Comparison between average values of LMA and LPP. MCRE 3

Speed (rpm)	Load (mbar)	Average LPP (CAD)	Average LMA (CAD)	Standard Error
2000	30	15.4	16	2.1
2000	50	27.2	29.9	2.8
2000	100	25.3	29.5	1.9
3000	30	12.5	11.3	0.8
3000	50	20.4	19.4	1.2
3000	100	21.3	20.7	1.5
4000	30	14.1	10.8	1.3
4000	50	21	20.9	1.4
4000	100	22.6	22.3	3.4
4500	30	16.7	12.4	0.6
4500	50	25.4	22.6	3.5
4500	100	25.6	23.4	4.6

using the LMA is listed too. The standard error value is similar for all the conditions and it is lower than 5. This result confirms that accelerometers can be exploited as non-intrusive sensors in a closed-loop control system that uses as the crank angle of the pressure peak as feedback value.

In Table 4.4, the average Maximum Pressure Rise (MPR) angular position, which is considered another important parameter for engine control, and the average LMA are compared for MCRE 3 engine. The table refers to the entire sets of tested engine operating conditions; the averages are calculated over 400 consecutive engine cycles. Also the standard errors, related to the estimation of MPR position by using LMA, are reported in the table. In particular, the maximum difference between average values of MPR position and of LMA is 0.2 CAD and the maximum standard error is 0.3 CAD for MCRE 3 facility.

In order to use the vibration signal for diagnosis and control of the Diesel combustion process (SCRE 1 and MCRE 3), the correlation between this signal and the in-cylinder mass fraction burned has been investigated. In particular, a strict correspondence between LMA and MFB50 (angular position where the 50% of the fuel mass inside the cylinder is burnt) can be found (see Fig. 4.5). MFB50, also called center of combustion, represents a very important combustion indicator, as widely described in the literature [14], and mentioned in the next chapter of the present book. Its value is strictly connected to the in-cylinder combustion progress and represents a stable measure of the timing of combustion. Moreover, in Diesel engines, a linear dependence of MFB50 on Start of Injection (keeping constant all

Table 4.4 Comparison between average values of LMA and MPR position. MCRE 3

Speed (rpm)	Average MPR (CAD)	Average LMA (CAD)	Standard error
2000	8.3	8.5	0.3
2000	11.3	11.2	0.3

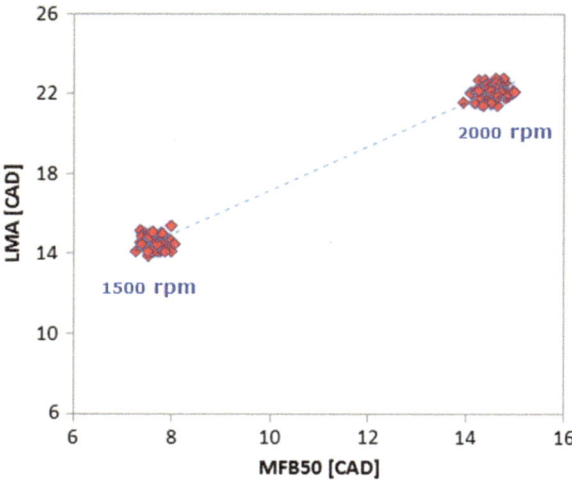

Fig. 4.5 Correlation between filtered vibration signal and in-cylinder mass fraction burned (LMA vs MFB50)

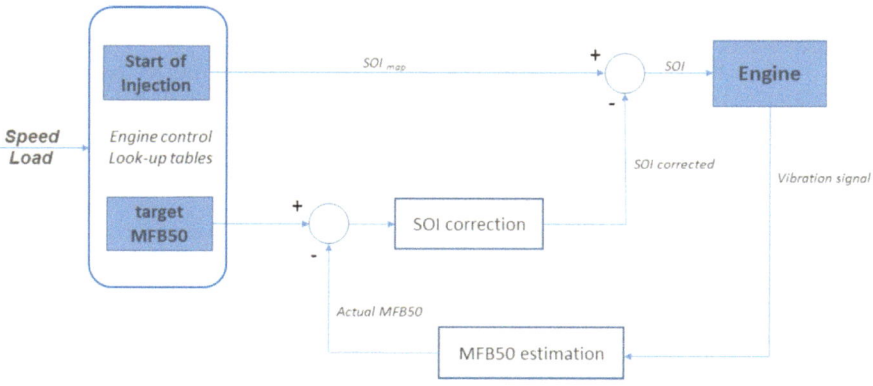

Fig. 4.6 Proposed architecture for real-time control of compression ignition engines via accelerometer signal

the other injection parameters and EGR rate) can be demonstrated [13–15]. Therefore, MFB50 can be used as a feedback variable in a closed-loop control system that acts on the Start of Injection (SOI) to obtain the optimal combustion timing in compression ignition engines.

A block scheme of such a closed-loop control is shown in Fig. 4.6. The goal of this controller is to maintain the actual MBF50 (estimated by means of the vibration signal) as close as possible to its optimal value. To this aim, SOI is real-time adjusted until the actual MFB50 value meets its target value stored in engine control maps.

Some criteria for use of MFB50 in closed-loop control architectures of SI engines are presented in the next chapter. On the other hand, it is well known that in

SI engines the optimal pressure peak location should be fairly constant (around 16 CAD ATDC), regardless of the operating conditions [14]. Therefore, a spark advance control algorithm able to maintain unchanged the LPP can be conceived. To this aim, the vibration signal could be used to estimate LPP. This information can be subsequently sent to the controller, which gives as output, on the basis of the LPP reference value (i.e., 16 CA ATDC), a new spark timing value. However, in the calculation of the new spark advance, the controller should also guarantee that the adjusted spark timing will not induce engine knock.

References

1. Mobley, C., Non-intrusive in-cylinder pressure measurement of internal combustion engines. SAE Technical Paper 1999-01-0544 (1999). doi:10.4271/1999-01-0544
2. R. Villarino, J. Böhme, Peak pressure position estimation from structure-borne sound. SAE Technical Paper 2005-01-0040 (2005), doi:10.4271/2005-01-0040
3. B. Badawi, M. Shahin, M. Kolosy, S. Shedied et al., Identification of diesel engine cycle events using measured surface vibration. SAE Technical Paper 2006-32-0097 (2006). doi:10. 4271/2006-32-0097
4. C. Polonowski, V. Mathur, J. Naber, J. Blough, Accelerometer based sensing of combustion in a high speed HPCR diesel engine. SAE Technical Paper 2007-01-0972 (2007). doi:10. 4271/2007-01-0972
5. J. Antoni, J. Daniere, F. Guillet, Effective vibration analysis of IC engines using cyclostationarity, part 1: a methodology for condition monitoring. J. Sound Vib. **257**, 815–837 (2002)
6. R. Johnsson, Cylinder pressure reconstruction based on complex radial basis function networks from vibration and speed signals. Mech. Syst. Signal Process. **20**, 1923–1940 (2006)
7. A.P. Carlucci, F.F. Chiara, D. Laforgia, Analysis of the relation between injection parameter variation and block vibration of an internal combustion diesel engine. J. Sound Vib. **295**, 141–164 (2006)
8. L. Barelli, G. Bidini, C. Buratti, R. Mariani, Diagnosis of internal combustion engine through vibration and acoustic non-intrusive measurements. Appl. Therm. Eng. **29**, 1707–1713 (2009)
9. O. Chiavola, G. Chiatti, L. Arnone and S. Manelli, Combustion characterization in diesel engine via block vibration analysis. SAE Technical Paper 2010-01-0168 (2010). doi:10.4271/2010-01-0168
10. L. Pruvost, Q. Leclere, E. Parizet, Diesel engine combustion and mechanical noise separation using an improved spectrofilter. Mech. Syst. Signal Process. **23**, 2072–2087 (2009)
11. M. El-Gharmy, J.A. Steel, R.L. Reuben, T.L. Fog, Indirect measurement of cylinder pressure from diesel engines using acoustic emission. Mech. Syst. Signal Process. **19**, 751–765 (2005)
12. F. Taglialatela, N. Cesario, M. Porto, S. Merola et al., Use of Accelerometers for spark advance control of SI engines. SAE Int. J. Engines **2**(1), 971–981 (2009). doi:10.4271/2009-01-1019
13. F. Taglialatela-Scafati, N. Cesario, M. Lavorgna, E. Mancaruso et al., Diagnosis and control of advanced diesel combustions using engine vibration signal, SAE Technical Paper 2011-01-1414 (2011) doi:10.4271/2011-01-1414
14. J.B. Heywood, *Internal Combustion Engine Fundamentals* (Mc Graw-Hill, NewYork, 1988)
15. H. Hülser, K. Neunteufl, C. Roduner, M. Weißbäck et al., EmIQ: Intelligent combustion and control for Tier2 Bin5 diesel engines. SAE Technical Paper 2006-01-1146 (2006) doi:10. 4271/2006-01-1146

Chapter 5
Use of in-Cylinder Pressure and Learning Circuits for Engine Modeling and Control

The parameter widely considered as the most important for the diagnosis of combustion process in internal combustion engines is the cylinder pressure. This signal represents, in fact, the most direct signal available for engine control [1] and numerous control algorithms based on pressure measurement as a feedback signal have been proposed.

However, measure of cylinder pressure can involve an intrusive approach to the cylinder and a special mounting process. Moreover, combustion pressure transducers used for these applications are very expensive, also because they are requested to resist to a highly hostile environment. For all these reasons, use of modeling and control techniques based on cylinder pressure analysis has been generally limited to research applications. However, more robust and cost-effective in-cylinder sensors have been recently developed and their usage in mass-produced vehicles now appears more feasible. On the other hand, use of a pressure sensor allows to replace many other sensors present in engines such as knock sensor and air mass meter. Typically, this new type of pressure sensors is integrated in the glow plugs [2], in the spark plugs, or into the injector valves.

Cylinder pressure signal includes most of the nonlinear phenomena occurring during engine combustion process. A typical example is represented by the so-called *cyclic variability* (also known as cyclic dispersion), which consists in a substantial variation from one cycle to the next in the combustion development. This phenomenon, which can lead to significant differences in the combustion efficiency from one engine cycle to another, is more evident in spark-ignited engines, especially when fueled with lean mixtures. Basically, cyclic variability is caused by variations in mixture motion within the cylinder at the time of the spark, variations in the amounts of air and fuel fed to the cylinder at each cycle, and variations in the mixing of fresh mixture and residual gases within the cylinder at each cycle [3]. Cyclic dispersion can be described as a highly nonlinear, or even chaotic process [4, 5], even if some researchers have highlighted its inherently non-deterministic and stochastic nature. The most evident effect of cyclic variation on cylinder pressure is a clear variability of pressure traces versus time from cycle-to-cycle (see Fig. 5.1).

© The Author(s) 2018   55
F. Taglialatela Scafati et al., *Nonlinear Systems and Circuits in Internal Combustion Engines*, SpringerBriefs in Nonlinear Circuits,
https://doi.org/10.1007/978-3-319-67140-6_5

Fig. 5.1 Example of cyclic
dispersion of pressure signal
in a SI engine

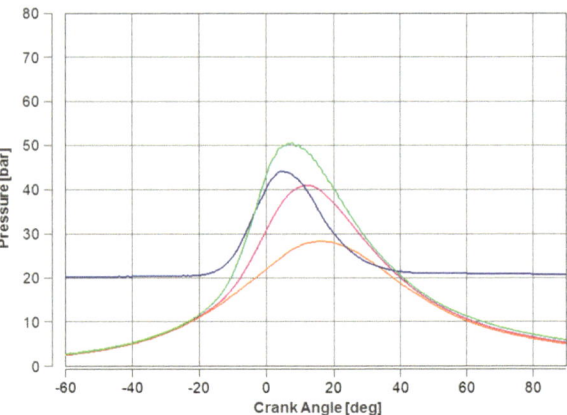

Use of real-time cylinder pressure in control architectures for both SI and Diesel engines offers a variety of significant advantages in terms of improved engine performances and reduced toxic emissions. Closed-loop controls using cylinder pressure measurements are able to compensate for manufacturing variations, aging and wear, fuel properties, and some environmental factors like humidity. In SI engines, some examples of applications that can achieve great benefits from the use of pressure signal are *knock detection and control*, *air–fuel ratio* control, which can be in this way enabled also during transient operation and cold-starts, *lean-burn* control, *fault diagnosis* (e.g., misfire detection), estimation *of cylinder air charge*, and *exhaust emission monitoring*. Moreover, cylinder-pressure-based feedback control allows to achieve an *individual-cylinder* control, in which optimal combustion control can be maintained in each cylinder of the engine. This permits to compensate for cylinder-to-cylinder A/F imbalances that can be due to both unequal fuel flow and unequal airflow to the cylinders.

In Diesel engines, closed-loop combustion control based on real-time combustion information coming from a cylinder pressure sensor is a key element that enables use of the so-called *alternative Diesel combustion*s (e.g., Homogeneous Charge Compression Ignition) in series production engines.

5.1   Cylinder Pressure Analysis and Extraction of Parameters for Engine Combustion Diagnosis and Control

Use of whole pressure trace for engine combustion diagnosis and control requires a high computational effort to modern ECU. For this reason, several algorithms that employ parameters derived from the pressure signal, and strictly related to the behavior of the combustion process, have been proposed for control-oriented applications. To this aim, *location of cylinder peak pressure* (LPP) has been suggested as a parameter for closed-loop control of spark timing in SI engines [6]. However, use of LPP as a unique combustion indicator in pressure-based control systems shows a reduced applicability, especially when used for diluted air–fuel mixtures, e.g., during idle or with high EGR rates [7]. Another method for analyzing cylinder pressure signal is based on the use of the so-called *momentum* of pressure trace (*vs* crank angle) as an input to an estimator algorithm. This approach was successfully applied for air–fuel ratio control [8], but the calculation procedure is quite computationally intensive [9] and, for this reason, it was not extended to other control functions.

Indicated Mean Effective Pressure (IMEP), which can be calculated from the cylinder pressure waveform during the combustion event, was used as another possible parameter for the diagnosis of combustion efficiency. In particular, several algorithms [10] for engine torque fluctuations monitoring on the basis of IMEP value have been developed. Moreover, this information has been used to identify the combustion dilution limits and to detect the occurrence of misfired cycles.

Other approaches proposed for real-time pressure-based control of internal combustion engines are the so-called *pressure ratio* and *difference pressure* methods.

The *pressure ratio (PR)* algorithm [9] involves the calculation of the ratio between the fired pressure and the corresponding motored pressure (P_{mot}) at each engine crank angle:

$$PR = \frac{P(\theta)}{P_{mot}(\theta)}.$$ (5.1)

The *PR* has a unity value before combustion and rises during combustion to a final pressure ratio value (*FPR*), which, depending on the amount of heat release per unit charge mass, usually ranges from 2.8 to 4 and typically occurs at around 55 crank angles ATDC for spark-ignited engines.

A strict correlation between the fuel mass fraction burned (*MFB*) and the pressure ratio exists. In fact, it can be thermodynamically demonstrated that

$$MFB(\theta) \approx PR(\theta) - 1.$$ (5.2)

Therefore, the fractional rise in the pressure ratio can be considered an estimate of the mass fraction burned, which only slightly is influenced by piston motion and heat transfer [9].

According to [9], pressure ratio calculated during combustion at 10° ATDC, normalized to the *FPR*, and indicated as *PRM10*, provides a very sensitive measure of combustion phasing and can thus be used as a variable in algorithms for closed-loop spark timing control. In [9], a spark control system was implemented by controlling the *PRM10* to a nominal value of 0.55, even if modifications to this target were made for some special operating conditions (e.g., for decelerations or during knock).

FPR reaches its maximum value for stoichiometric mixtures and decreases as the mixture dilution increases; therefore, it can be a useful indicator of the total charge dilution, and it is applicable to the control of lean-burn systems, of systems that use high amount of EGR or of systems that vary the amount of residuals through variable valve trains [9].

Another method for combustion diagnosis and control is the so called *difference pressure* method, which uses the fired cylinder pressure signal, subtracted by the corresponding motored pressure, as a control parameter.

$$\Delta P(\theta) = P(\theta) - P_{\text{mot}}(\theta). \tag{5.3}$$

This approach, according to [11], allows to solve the problem of a reliable absolute pressure measurement without drifting and scaling issues, which is required when the *pressure ratio* method is used.

Several pressure-based algorithms using the *difference pressure* approach have been proposed. Most of these employ parameters extracted from the difference pressure curve, such as the difference pressure curve peak $(\theta_{\text{peak}}, \Delta P_{\text{peak}})$ or the length of a secant to the difference curve at some specific crank angles.

However, it has to be pointed out that both *pressure ratio* and *difference pressure* approaches require an approximation of the motored pressure curve, which can be reconstructed using a polytropic approximation.

One of the most used parameters in pressure-based control algorithms is the position of the 50% of burn mass (MFB50), i.e., the crank angle at which 50% of the fuel is converted. MFB50 is considered an important combustion indicator; its value is strictly connected to the in-cylinder combustion progress and represents a stable measure of the timing of combustion. MFB50 value affects the engine thermal efficiency, peak cycle temperature and pressure, and exhaust emissions. It can be calculated from the pressure signal using for example Eq. (5.2), in which MFB value can be estimated quite easily on the basis of the pressure ratio. A linear dependence of MFB50 on Start of Injection (SOI), keeping constant all the other injection parameters and EGR rate, has been demonstrated in Diesel engines. [12]. On the basis of this feature, MFB50 can be used as a feedback variable in Diesel engine closed-loop control systems acting on SOI to obtain optimal combustion timing (more details are in the previous chapter).

In [13], it has been shown, on the basis of thermodynamic considerations and experimental results, that a MFB50 value of approximately 8–10 CA degrees after TDC guarantees an optimal combustion efficiency in SI engines, irrespective of the considered engine operating condition. This feature has been used in many pressure-based ignition control architectures in SI engines. Some examples will be shown in the next paragraphs.

5.2   Use of Learning Algorithms in Pressure-Based Engine Controls

The relation between the value of most of the above indicated parameters extracted from the pressure curve and engine output signals, such as engine torque, exhaust emissions is often highly nonlinear. Therefore, engine control architectures based on the analysis of the pressure signal generally require the modeling of these correlations. To this aim, learning algorithms, and in particular neural networks, can be employed for their feature of being valuable tools for approximation and control of nonlinear dynamic systems. Neural networks are also used in real-time adaptive controllers in order to enhance the feed-forward control action during engine transients. An example of this can be found in [11], where an adaptive neural network controller is employed in a pressure-based ignition control system. In general, the objective of an ignition control is to maximize the engine efficiency for each engine operating condition, avoiding at same time knocking phenomena. This goal, as mentioned before, can be obtained if the ignition point is set such that at a specified crank angle (9 CA after TDC) exactly 50% of fuel is burned. Therefore, in a closed-loop ignition control system, the condition MFB (9 CA) = 0.5, can be used as a control objective. However, such a feedback control leads to good performance only under steady state and slowly time variants conditions. This is due to the fact that, based on the measurement from the current engine cycle, optimal ignition time can only be computed for the next cycle. Thus, a dead time of one cycle exists. Moreover, as severe cycle-to-cycle fluctuations can be found even under steady state conditions, the results of cylinder pressure evaluation have to be averaged over several engine cycles (at least 10 cycles). This further slows down the control action during transients. For this reason, in [11], the control system is enhanced by adding an adaptive neural feedforward controller for each cylinder, with online learning action performed during the normal operation of the engine.

5.3   Use of Combustion Pressure Signal for Cylinder Air Charge Estimation

Three-way catalytic converters (TWC) allow spark-ignited engines to obtain significant reduction of exhaust emissions. A precise control of air–fuel ratio (AFR) to the stoichiometric value is necessary to achieve the maximum efficiency of TWC in the conversion of the toxic exhaust gases (CO, NO_x, HC) into less harmful products (CO_2, H_2O, N_2).

Traditional AFR control strategies are composed by a feed-forward part, in which the amount of fuel to be injected is calculated on the basis of the in-cylinder mass airflow, and a feedback part, which uses the signal of an oxygen sensor (lambda sensor) placed at the exhaust to ensure that AFR remains in a strict neighborhood of stoichiometric value. Generally, the feedback part of an AFR control system is fully active only in steady state conditions; moreover, the oxygen sensor signal is available only after that this sensor has reached a fixed operating temperature. Thus, during transients and cold starting, the feedback control is deactivated and the feedforward component of AFR control assumes a greater significance. Estimation of cylinder air charge is the basis for calculating the amount of fuel to inject in the feedforward part of an AFR control system.

A conventional technique [3] for estimating the cylinder air charge in a spark-ignited engine involves the use of a *speed–density* equation:

$$\dot{m}_{\text{ap}} = \eta\,(p_m, N)\,\frac{V_d N}{120}\frac{p_a}{R T_m}, \tag{5.4}$$

where $\dot{m}_{\text{ap}}$ is the inlet air mass flow rate, V_d is the engine displacement, N is the engine angular speed, and p_a is the ambient pressure. T_m and p_m are the mean manifold temperature and pressure, respectively, whereas η is the engine volumetric efficiency. This latter can be considered a highly nonlinear function of engine speed (N) and manifold pressure (p_m). It can be calculated via experimental for different engine operating points. A standard method is to map the volumetric efficiency and compensate it for density variations in the intake manifold. One of disadvantages in using the *speed–density* equation for the in-cylinder airflow estimation is the uncertainty in the volumetric efficiency. Generally, the volumetric efficiency is calculated in the calibration phase with the engine under steady state conditions. However, variations in the volumetric efficiency due for example to engine aging and wear, buildup of deposits in the combustion chamber, etc., can induce errors in the air charge estimation. Moreover, the low-pass characteristic of the commercial *Manifold Absolute Pressure* (MAP) sensors used for the determination of the manifold pressure makes the signal affected by a delay that introduces an error in the calculation of cylinder air charge during the fast transients. This problem is not solved by using a faster sensor: in this case the sensor captures also pressure fluctuations due to the valve and piston motion [14].

It has also to be considered that in engines equipped with an EGR valve, the MAP sensor cannot distinguish between fresh air and inert exhaust gas in the intake manifold. Therefore, in this case, the *speed–density* equation cannot be used as written in (5.4), and the air charge estimation algorithm must provide a method for separating the contribution of recirculated exhaust gas to the total pressure in the intake manifold [15].

An alternative method for the air charge determination is to use a mass airflow (MAF) sensor located upstream from the throttle body, which measures directly the inlet airflow. The main advantages of a direct airflow measurement are [3]:

– automatic compensation for engine aging and for all factors that modify engine volumetric efficiency;
– improved idling stability;
– lack of sensibility of the system to EGR, since only fresh airflow is measured.

Anyway, airflow measurement by means of a MAF sensor, which generally is a hot wire anemometer, accurately estimates the flow into the cylinder only in steady state, while in transients the intake manifold filling/emptying dynamics play a significant role [16, 17]. In addition, MAF sensors for commercial automotive applications have a relatively high cost compared to the cost of the MAP sensors used in the *speed-density* approach.

An alternative way to estimate engine air mass charge involves the use of cylinder pressure measurement. This approach, which assumes each cylinder to be equipped with a pressure sensor, offers the possibility of a cylinder-individual control, resulting in a more uniform load and air/fuel ratio distribution. Moreover, even though use of combustion pressure sensors can represent a high cost, especially for low-end cars, it allows to totally replace other types of air mass sensors.

A neural network model, which uses the signal from a low-cost combustion pressure transducer for prediction of cylinder air charge, is presented in [18]. Basically, the heart of the system is a *mass airflow estimator,* which is a model able to predict in real-time the air charge as a function of some features extracted from the combustion pressure signal. The output of the MAF estimator can then be used by the engine control maps, in the feedforward part of an air/fuel control system, to calculate the optimal injection time for a stoichiometric intake mixture.

The MAF estimator has the following inputs:

- *throttle position signal*, coming from a dedicated sensor used to monitor the throttle opening;
- *engine angular speed*, typically evaluated by means of a variable reluctance sensor;
- *engine angular absolute position;*.
- *cylinder pressure signal*, coming from a combustion pressure transducer.

The output is represented by an estimated mass airflow value, expressed as kg/h.

From a mathematical point of view, the MAF estimator can be described as a learning machine trained in order to emulate the way of functioning of a common MAF sensor used in the traditional approaches for cylinder air charge estimation. A MLP neural network was used as learning machine model. The choice of all the endogenous parameters of the neural network model, such as the number of hidden layer, the number of neurons for each hidden layer, the type of activation functions, the value of the regularization parameter was made in order to maximize the "generalized forecast capability" of the learning machine.

Mass airflow estimator included some sub-blocks devoted to the processing of neural network input and output signals. By means of an input data preprocessing, they were transformed in dimensionless data with a zero mean and a unitary variance. Moreover, on these new dimensionless data a Principal Component Analysis (PCA) transform was carried out. A post-processing sub-block performed the inverse of the previous transforms. Lastly, an *Edge Detector* sub-block was used to enable or disable the model prediction according to engine angular position value. In particular, airflow estimation is enabled only in the crank angle interval when the air induction process is realized. Therefore, the Edge Detector enables the model functionality when the engine is at TDC at the end of the exhaust stroke (i.e., when the intake stroke is starting). Then, model prediction is disabled when inlet and exhaust valves are simultaneously closed (at 140 crank angles BTDC in the compression stroke for the test engine used in this study). This event occurs when inlet air charge reaches its maximum value.

The mass airflow estimated by the model can be used for the calculation of fuel mass amount to deliver during the subsequent engine cycle. Therefore, an inherent delay, which assumes more importance during transient operations, is introduced with this approach.

The neural network was trained on an experimental data set including different engine speeds and loads. In particular, three engine speeds (3000, 4600, and 5800 rpm) and eight throttle positions corresponding to different engine loads were considered. For each engine operating condition, the respective air mass flow value was acquired by means of a hot wire anemometer mounted upstream of the intake manifold. The engine used both for the training of the mass airflow estimator and for its validation was a four stroke single-cylinder SI engine with a displacement of 125 cm^3.

In the following figures, the comparison between the mass airflow value (measured by means of a MAF sensor) and the mass airflow estimated by the soft computing model is shown. The figures refer to the engine in steady state conditions. In particular, in Figs. 5.2 and 5.3, the measured and the estimated mass airflows are compared for an engine speed of 4600 and a throttle opening of 70 and 20%, respectively. Figures 5.4 and 5.5 refer to an engine speed of 5800 rpm and a throttle opening of 30 and 100% (wide open throttle condition).

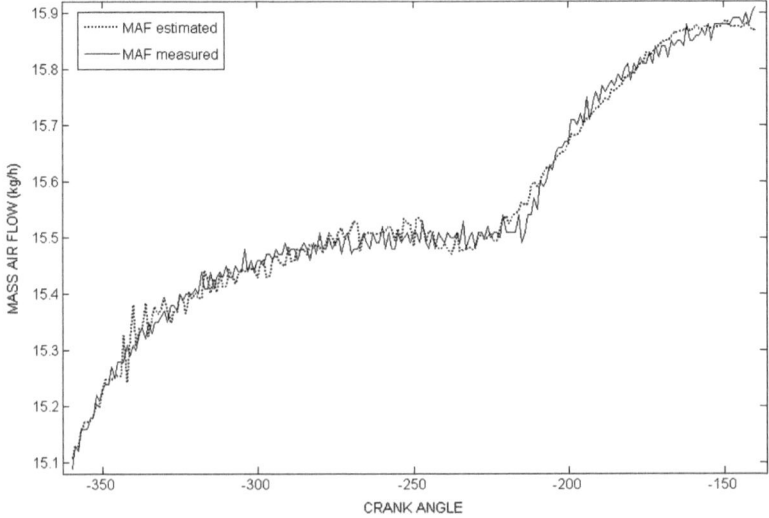

Fig. 5.2 Comparison between the estimated and measured MAF at 4600 rpm and 70% of throttle opening

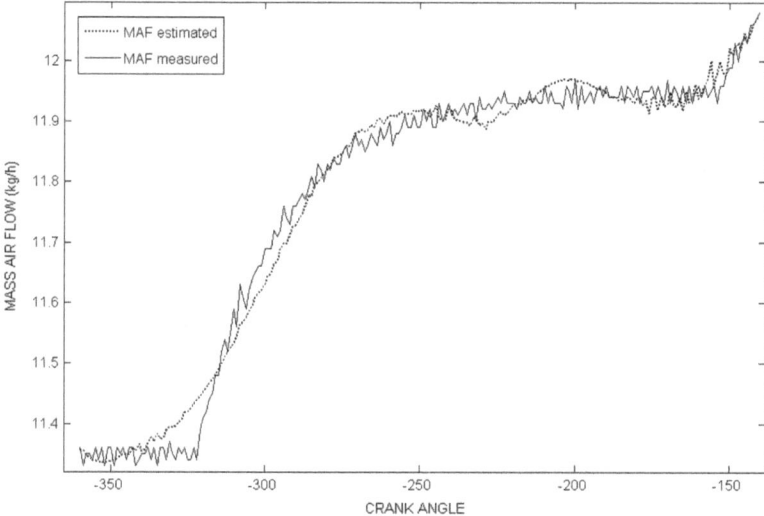

Fig. 5.3 Comparison between the estimated and measured MAF at 4600 rpm and 20% of throttle opening

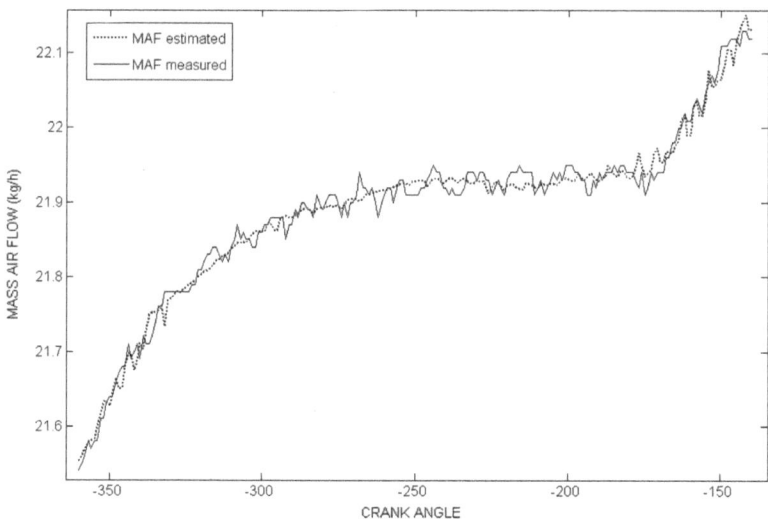

Fig. 5.4 Comparison between the estimated and measured MAF at 5800 rpm and wide open throttle condition

The results of the study confirmed that the pressure-based MAF model is capable of providing a good estimation of the intake air in steady state, with a mean square error lower than 3% for all the investigated engine operating conditions (see Table 5.1).

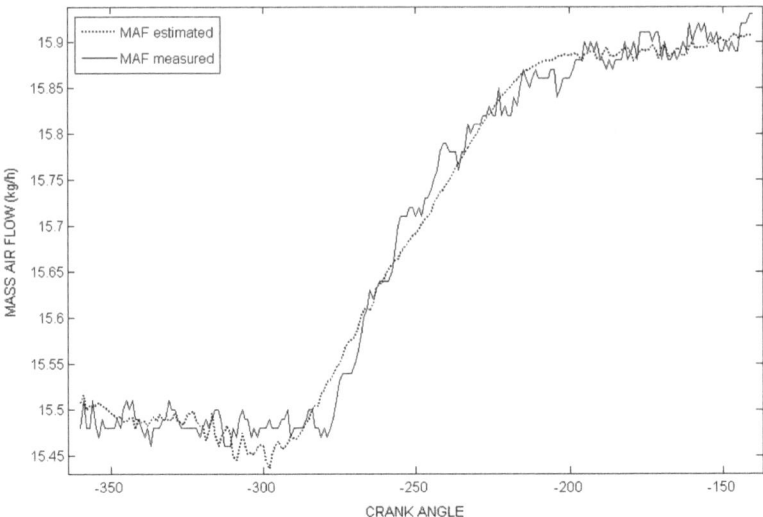

Fig. 5.5 Comparison between the estimated and measured MAF at 5800 rpm and 30% of throttle opening

Table 5.1 Mean square error in mass airflow estimation

Speed (rpm)	Throttle opening (%)	Mean square error
4600	70	0.011
4600	20	0.029
5800	100	0.015
5800	30	0.010

As mentioned before, the intake manifold filling/emptying dynamics play a dominant role in cylinder air charge estimation during transient conditions. However, using in-cylinder pressure curve as an input, cylinder air estimation is less sensitive to these dynamics and a better prediction of the intake air value is generally provided. In any case, due to its lack of precision in transients, the measure provided by an anemometer mounted upstream of the intake manifold cannot be employed for the validation of pressure-based models during transients.

5.4   Engine Knock Detection and Control Using in-Cylinder Pressure Measurement

In spark-ignited engines, knock is an abnormal combustion that occurs when the temperature and the pressure of the unburned gas in the cylinder increase too much, causing the fuel to self-ignite: this results in an oscillating pressure wave in the combustion chamber. A common reason for engine knock is the use of poor quality gasoline with low octane rating and the tendency to ignite prematurely. The octane-number requirement of an engine depends on how its design and the conditions under which it is operated affect the temperature and pressure of the end-gas ahead of the flame and the time required to burn the cylinder charge [3]. In spark ignition engines, knock is a phenomenon that limits how efficiently the engine can operate: the high oscillating pressure can cause damages, which can lead to a shorter lifetime for the engine. Severe knock can damage the engine even after only one or few self-ignitions. For these reasons, the detection of knock and the evaluation of its intensity are critical issues for electronic engine control systems and several methods have been proposed and used for this purpose [19, 20]. Most of the techniques for knock detection and control currently employed on mass-produced engines are based on engine block vibration analysis [21, 22]. The high-frequency pressure oscillations inside the cylinder caused by knocking events are transmitted to the engine block in the form of vibrations. The engine block vibration signal is measured using one or more accelerometers, often tuned to the nominal first resonance mode of the engine, mounted on the engine block. The accelerometers give a signal that is rectified using a pass-band filter. The values of total signal energy over time or of the maximum amplitude of the filtered signal are used as knock indicators and are compared to a knock discriminating threshold value. One advantage of using accelerometers for knock detection is that, with careful placement, only one or two sensors are required to monitor all cylinders. In addition, these sensors are relatively cheap. Anyway using this approach, it is important to distinguish knock-induced vibrations from block vibrations which can occur even during normal combustion. These vibrations are mainly induced from valve closing events and piston slaps, and they have a level that depends on the operating state of the engine (especially engine speed). Therefore, the effectiveness of the block vibration method for the knock detection is influenced by the appropriate choice of the threshold values, and it is generally a method that gives a poor signal-to-noise ratio (SNR) at high engine speeds due the high level of background noise. The most sensitive and reliable method for knock detection involves direct pressure measurement in each cylinder of the engine, since this measurement is not measurably affected by other mechanical sources. The necessity of one sensor for each cylinder and the high cost of the traditional pressure transducers limit the use of detection methods based on cylinder pressure analysis to research applications. However, the recent development of low-cost pressure sensors for EMS applications will make feasible in the next future the use of pressure transducers for knock detection and control for mass-produced engines.

As mentioned before, high-pressure oscillations, whose amplitude decays with time, can be observed on the in-cylinder pressure signal during knock. The amplitude of the pressure oscillations depends on the amount of the end-gas that undergoes self-ignition, and therefore, it inherently depends on the knock intensity. Moreover, the pressure waves have a characteristic frequency that is influenced by the characteristic length of the oscillation and by the speed of the sound in the combustion chamber [23], but it does not depend on engine angular speed. Figure 5.6 shows the typical oscillations occurring on the pressure curve during a knocking combustion.

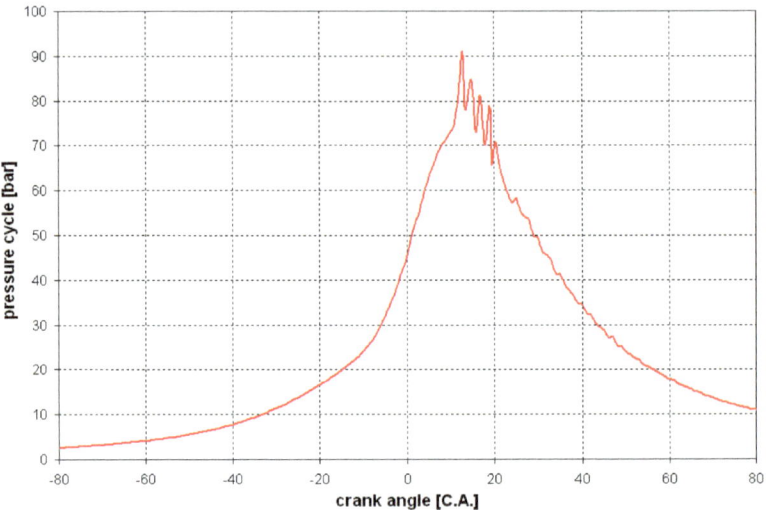

Fig. 5.6 High-frequency pressure fluctuations on the pressure cycle due to knock

A proper band-pass filtering of the pressure curve allows to separate the superimposed knock pressure waves from the cylinder pressure bias. Several parameters of the pass-band filtered pressure signal have been proposed as knock indicators. The simplest knock indicator is represented by the maximum amplitude of band-pass filtered data [24]. Many other knock indicators from the pressure signal have been suggested, such as the integral of the absolute value of pressure oscillations or the integral of the absolute value of the first derivative of pressure oscillations [25, 26]. However, use of these indicators makes the knock detection a very compute intensive task for the standard microcontroller architectures used in the automotive field.

In [27], a system using in-cylinder pressure curve and soft computing techniques for knock detection and control is presented. According to this method, the high-pressure oscillations are extracted from the cylinder pressure curve using a band-pass filtering. Then, the filtered curve is digitalized by comparison with a fixed threshold, and a variable number of digital pulses are produced. The number and the total duration of these digital pulses, by using fuzzy logic techniques, provide information about the presence of knock and its intensity in the form of a *knock index*. A knock controller will, then, use this latter index to modify the output of engine control maps delaying the ignition angle until the knock disappears.

Fuzzy logic techniques are traditionally used for the processing of rough and qualitative data, which can be explained by qualitative terms of the human language (e.g., "high temperature," "medium temperature," "low temperature," etc.) [28, 29]. The estimation of knock intensity can certainly be seen as qualitative information that can be evaluated using fuzzy algorithms.

Compared to traditional systems, a fuzzy logic approach for knock detection allows to find high-quality nonlinear correlations between inputs (knock indicators) and outputs (knock intensity level). Moreover, knock estimation by using fuzzy algorithms is easy to implement and requires a low computational time.

As mentioned before, the knock estimator presented in [27] uses as knock estimator the number and the total duration of the digital pulses obtained from the digitalization of the filtered pressure curve.

In particular, the following fuzzy sets of the variable *Pul_Cnt* (number of digital pulses) are considered:

- *npulse-abs*: *"absence of digital pulses"*
- *npulse-low*: *"low number of digital pulses"*
- *npulse-medium*: *"medium number of digital pulses"*
- *npulse-high*: *"high number of digital pulses"*

with the corresponding membership functions indicated in Fig. 5.7.

For the variable "total duration of digital pulses" (*Pul_Dur*), the following fuzzy sets are considered:

- *tpulst-abs* *"total pulse duration zero"*
- *tpulst-low* *"total pulse duration low"*
- *tpulst-medium* *"total pulse duration medium."*
- *tpulst-high* *"total pulse duration high"*

with the corresponding membership functions depicted in Fig. 5.8.

The knock estimator implements a fuzzy algorithm with *Pul_Cnt* and *Pul_Dur* as antecedents and as consequents; therefore, the fuzzy sets are defined as follow (Fig. 5.9):

- *Kgrade-abs* *"Knock absence"*
- *Kgrade-low* *"Light knock"*
- *Kgrade-medium* *"Medium knock"*
- *Kgrade-high* *"Heavy knock"*

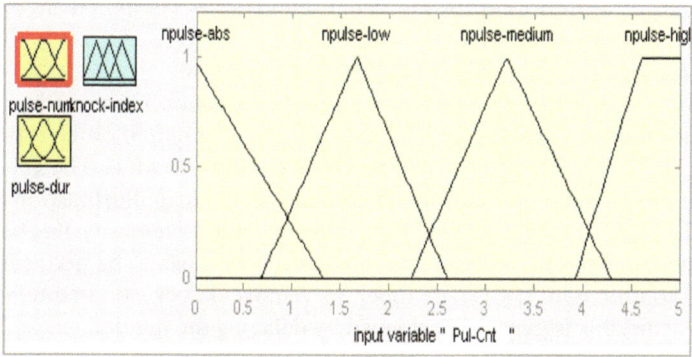

Fig. 5.7 Fuzzy sets of the variable *Pul_Cnt* (number of digital pulses) and their membership functions

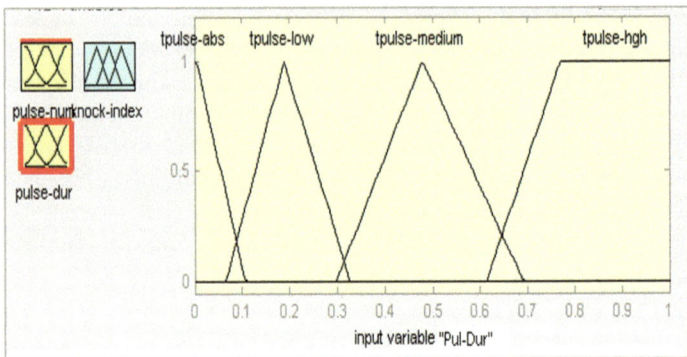

Fig. 5.8 Fuzzy sets of the variable *Pul_Dur* (total duration of digital pulses) and their membership functions

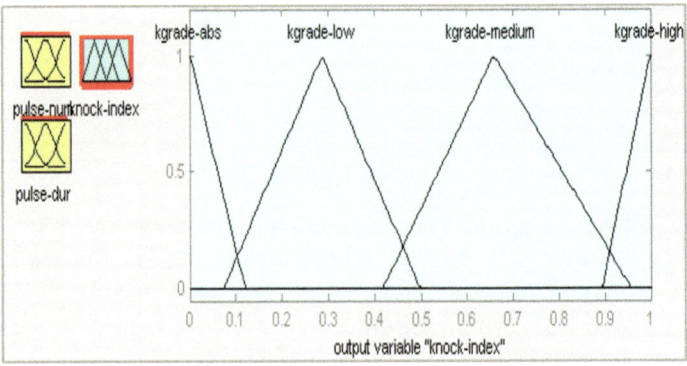

Fig. 5.9 Fuzzy sets of the variable Knock Index and their membership functions

The fuzzy algorithm used for the knock estimator is composed by eight rules, whose general expression is

IF npulse-low AND tpulst-medium THEN Kgrade-low.

The output of the fuzzy algorithm with de-fuzzification is shown in Fig. 5.10.

For each couple of inputs *Pul_Cnt* and *Pul_Dur*, the output of the estimator is a *Knock Index* value, which is a number ranging from 0 to 1. The input–output correlation can be represented by the three-dimensional map illustrated in Fig. 5.11.

Once the knock occurrence has been detected and its intensity has been calculated by means of the fuzzy estimator, a control action has to be performed by the electronic control unit (ECU) in order to remove knock. In traditional engine control systems, this is generally achieved by delaying the ignition angle. This latter is the fastest control action because changes take effect already in the next cycle.

The control system can modify the value of ignition through an instruction of the following type:

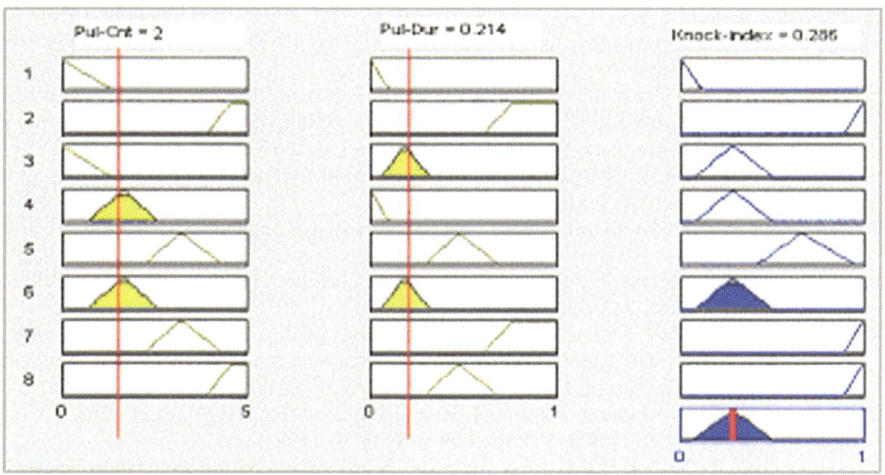

Fig. 5.10 Values of Knock Index for each couple of inputs *(Pul_Cnt and Pul_Dur)*

Fig. 5.11 Three-dimensional
map of knock index

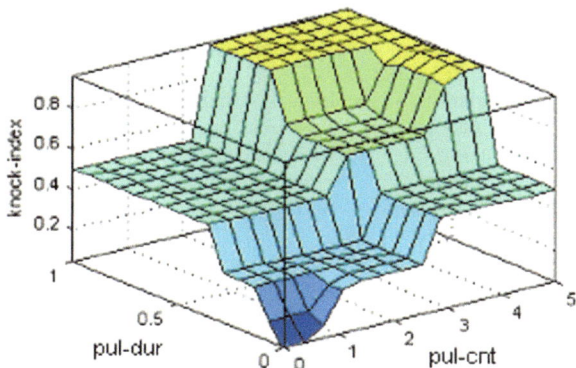

$$\text{Ignition Angle} = (\text{Ignition Angle}_{\text{Map}}) + K * (\text{Knock Index}),$$

where K is a proportionality factor, which determines how gradual the control
action has to be.

References

1. L. Guzzella, C. Onder, *Introduction to modeling and control of internal combustion engine systems* (Springer Science & Business Media, 2009)
2. M.T. Wlodarczyk, High accuracy glow plug-integrated cylinder pressure sensor for closed loop engine control. SAE Technical Paper no. 2006-01-0184 (2006)
3. J.B. Heywood, *Internal Combustion Engine Fundamentals* (McGraw-Hill, 1988)

4. C. S. Daw, C.E.A. Finney, J.B. Green, M.B. Kennel, J.F. Thomas, F.T. Connolly, A simple model for cyclic variations in a spark-ignition engine. SAE Technical Paper no. 962086 (1996)
5. C. Letellier, S. Meunier-Guttin-Cluzel, G. Gouesbet, F. Neveu, T. Duverger, B. Cousyn, Use of the nonlinear dynamical system theory to study cycle-to-cycle variations from spark ignition engine pressure data. SAE Technical Paper no. 971640 (1997)
6. I. Glaser, J.D. Powell, Optimal closed-loop spark control of an automotive engine. SAE Technical Paper no. 810058 (1981)
7. F.A. Matekunas. Modes and measures of cyclic combustion variability. SAE Technical Paper no. 830337 (1983)
8. J.D. Powell, Engine control using cylinder pressure: past, present, and future. J. Dyn. Syst. Meas. Contr. **115**(2B), 343–350 (1993)
9. M.C. Sellnau, F.A. Matekunas, P.A. Battiston, C.F. Chang, D.R. Lancaster, Cylinder-pressure-based engine control using pressure-ratio-management and low-cost non-intrusive cylinder pressure sensors. SAE Technical paper no. 2000-01-0932 (2000)
10. N. Kobayashi, T. Akatsuka, J. Nakano, T. Kamo, S. Matsushita, Development of the Toyota lean combustion system. Int. J. Veh. Des. **5**(6), 731–738 (1984)
11. S. Leonhardt, N. Muller, R. Isermann, Methods for engine supervision and control based on cylinder pressure information. IEEE/ASME Trans. Mechatron. **4**(3), 235–245 (1999)
12. H. Hulser, K. Neunteufl, C. Roduner, M. Weissback, L. Burgler, M. Glensvig, EmQI: intelligent Combustion and Control for Tier2 Bin5 Diesel Engines. SAE paper no. 2006-01-1146 (2006)
13. M. Bargende, Most optimal location of 50% mass fraction burned and automatic knock detection. Components for automatic optimization of SI-engine calibrations. MTZ Worldwide **56**(10), 632–638 (1995)
14. O. Barbarisi, A. Di Gaeta, L. Glielmo, S. Santini, An extended Kalman observer for the in-cylinder air mass flow estimation, in *MECA02 International Workshop on Diagnostics in Automotive Engines and Vehicles* (2001)
15. M. Jankovic, S.W. Magner, Air charge estimation and prediction in spark ignition internal combustion engines, in *Proceedings of the American Control Conference*, San Diego, California, June (1999)
16. J.W. Grizzle, J.A. Cookyand, W.P. Milam, Improved cylinder air charge estimation for transient air fuel ratio control, in *Proceedings of American Control Conference* (1994)
17. I. Stotsky, A. Kolmanovsky, Application of input estimation and control in automotive engines. Control Eng. Pract. **10**, 1371–1383 (2002)
18. F. Taglialatela, N. Cesario, M. Lavorgna, Soft computing mass air flow estimator for a single-cylinder SI engine. SAE Technical Paper no. 2006-01-0010 (2006)
19. F. Millo, C.V. Ferraro, Knock in S.I. engines: a comparison between different techniques for detection and control. SAE Paper 982477 (1998)
20. K. Schmillen, M. Rechs, Different methods of knock detection and knock control. SAE Paper 910858 (1991)
21. J. Lee, S. Hwang, J. Lim, D. Jeon, Y. Cho, A new knock-detection method using cylinder pressure, block vibration and sound pressure signals from a SI engine. SAE Paper 981436 (1998)
22. K. Chun, K. Kim, Measurement and analysis of knock in a SI engine using the cylinder pressure and block vibration signals. SAE Paper 940146 (1994)
23. C. Elmqvist, F. Lindström, H.E. Ångström, B. Grandin, G. Kalghatgi, Optimizing engine concepts by using a simple model for knock prediction. SAE Technical Paper 2003-01-3123 (2003)

24. W.R. Leppard, Individual-cylinder knock occurrence and intensity in multicylinder engines. SAE Paper 820074 (1982)
25. P.V. Puzinauskas, Examinations of methods used to characterize engine knock. SAE Paper no. 920808 (1992)
26. E. Antonelli, Definizione e misura dell'intensità di detonazione. ATA-July (1967)
27. F. Taglialatela, G. Moselli, M. Lavorgna, Engine knock detection and control using in-cylinder pressure signal and soft computing techniques. SAE Technical Paper no. 2005-24-061 (2005)
28. M. Lavorgna, M. Lo Presti, G. Rizzotto, Me todologie per la sintesi e l' analisi dei controllori fuzzy. Cavallotto edizioni (1996)
29. L. Fortuna, M. Lavorgna et al., Soft computing e valenze applicative. Cavallotto edizioni (1999)

Chapter 6
Identification and Compensation of Nonlinear Phenomena in Gasoline Direct Injection Process

Future emission regulations require the development of gasoline combustion engines with improved efficiency in order to obtain a strong reduction of the toxic emissions coupled to the reduction of fuel consumption and hence carbon dioxide emissions. The greatest fuel consumption benefit is achieved by means of systems such as the gasoline direct injection (GDI) combustion with unthrottled lean stratified operation. In this mode, the fuel is injected later in the compression stroke [1] allowing stable combustion of ultralean mixtures. However, the use of gasoline stratified charges can lead to several problems. In particular, due to the oxygen excess in the combustion stroke, the NO_x emission levels are generally higher than in the port fuel injection (PFI) engine or in homogeneous charge direct injection. Moreover, short time for mixture preparation and spray wall impingements are responsible for a high cycle-to-cycle variability and high particle emissions. On the other hand, the reduction of the particulate at the exhaust of gasoline direct engines represents a crucial aspect also considering the introduction of EU6 emission legislation that strongly pushes toward a reduction of the particulate emitted by the engines.

A potentially effective way to mitigate the problems of GDI stratified operation and reduce the wall impingement is the use of multiple fuel injections, splitting up the total fuel injection into several smaller (and shorter in duration) shots. The first effect of this approach is the reduction of the jet penetration into the combustion chamber, thus reducing the wall wetting and decreasing the particulate formation. Moreover, it has been demonstrated [2] that the use of multiple injection strategies allows to retard the angle of 50% mass fraction burned (MBF50), placing it close to the thermodynamic optimum. This allows to have lower combustion peak temperatures with a reduction of NO_x emissions.

However, this is not easy to obtain by the traditional GDI solenoid injectors. Indeed, the management of small injections forces GDI solenoid injectors to work in their so-called ballistic mode. The ballistic behavior appears at small injection pulse width when the pulse is cutoff before the valve fully lifts up. During the ballistic operation, the correlation between the electrical command and the injected

© The Author(s) 2018
F. Taglialatela Scafati et al., *Nonlinear Systems and Circuits in Internal Combustion Engines*, SpringerBriefs in Nonlinear Circuits,
https://doi.org/10.1007/978-3-319-67140-6_6

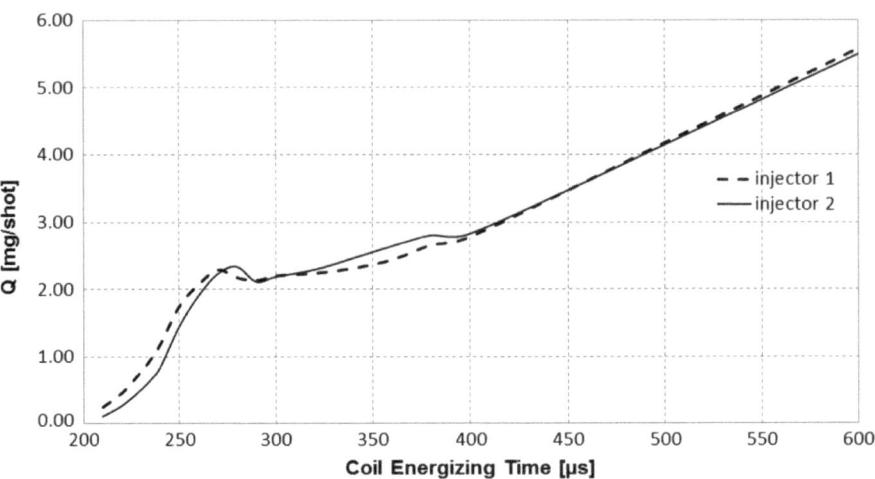

Fig. 6.1 Characteristic curves of the injection rate obtained for two injectors belonging to the same family

amount of fuel is highly nonlinear, the valve motion is unstable, and the fuel delivering cannot be controlled with optimum precision.

Figure 6.1 shows the characteristic curves of injection rate obtained for two different injectors belonging to the same family. The graphs represent an example of the typical dispersion. For both the injectors, in fact, the correlation between the pulse command duration and the injected fuel amount appears linear and repetitive for pulse durations greater than 400 µs and for injected fuel greater than 2.8 mg. For command pulse widths lower than 400 µs, the correlation becomes highly nonlinear and, for some pulse durations, may also invert the trend with a decrease of fuel quantity at the increase of the electrical command. In the nonlinear region, for very short injection command pulses, the injector needle starts the closure before having reached the widest lift position. It has been reported [3] that the causes of the nonlinear behavior mainly rely on the inertia of the injector spring–mass system and the reduction of the electromagnetic forces exerted by the coil, friction variations, and more. All these have an unpredictable effect on the dynamic of the needle lift due to manufacturing tolerances and aging effects, too. This is the reason for which, up to now, the nonlinear region has not been used in the traditional injection strategies for commercial GDI engines.

In order to achieve the desired injection target also during the ballistic operation, and to extend the use of solenoid injectors to short injections, real-time information about the actual fuel amount delivered at ballistic is needed. On the basis of this information, the injector energizing time can then be adjusted in real time by means of a closed-loop algorithm to obtain the requested injected quantity.

The graph in Fig. 6.2 illustrates a comparison between the injector voltage signal and the corresponding injected mass flow rate. The analysis of the two curves depicts that the voltage signal shows an inflection at the same time when the

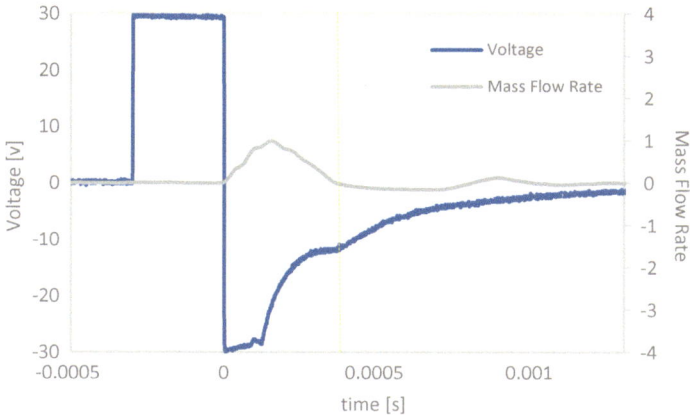

Fig. 6.2 Comparison between differential voltage command signal and the corresponding injected mass flow rate, for a coil energizing time of 300 μs

injected mass flow rate is annulled. This feature is not dependent on the value of coil energizing time.

The inflection on the voltage signal occurs during the switch-off phase, when the injector coil is de-energized and a self-induction voltage is created. In this phase, the voltage signal typically includes, other than a contribution due to decaying eddy currents, a contribution due to the movement of the needle in the de-energized coil, which has an amplitude depending on the needle speed. As this speed reaches its maximum value directly before the needle closing time, an inflection on the voltage signal appears at this time. On the basis of this feature, it comes out that a proper processing of the injector voltage signal allows to have real-time information about the needle closing time and then, in turn, on the actual amount of injected fuel. This latter can then be compared to absolute target values, and, as a result, the length of the electrical command can be consequently adjusted in real time during the ballistic operation mode. Using this approach, a closed-loop control architecture of injection duration in GDI engines can be introduced (see the block scheme in Fig. 6.3). This control architecture allows to manage the delivery of small fuel amounts in order to increase the minimum fuel injection capabilities of GDI solenoid injectors and permits to extend their use to the multiple fuel injection strategies.

As indicated in Fig. 6.3, the voltage signal in the de-energized coil has to be firstly filtered by means of a low-pass filter to reduce the superimposed noise. On the basis of results of a spectral analysis, a low-pass moving average filter with 40 coefficients and a cutoff frequency of 12 kHz has been proposed in [4]. The filtered signal is subsequently compared with a reference voltage signal (V_{ref}). The reference signal V_{ref} represents the voltage induced in the de-energized coil using an electrical command that does not produce any movement of the injector needle (e.g., due to a too short command pulse). Therefore, this latter signal exclusively is

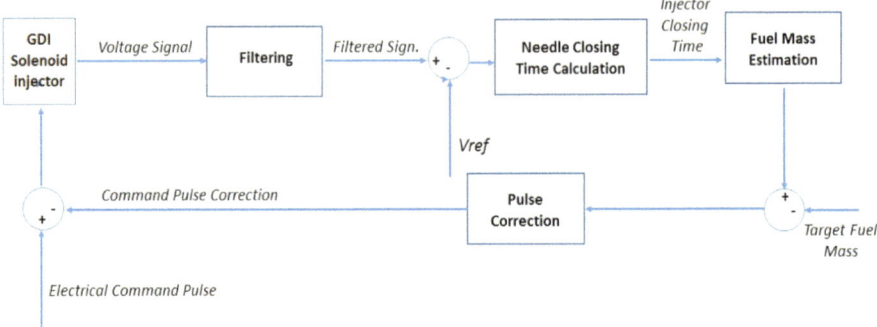

Fig. 6.3 Closed-loop control algorithm for real-time compensation of fuel quantity injected during the ballistic operation

related to the self-induced voltage due to the decaying eddy currents and it does not include the above-mentioned component due to the needle motion. As a consequence, the difference between the voltage signal and the reference signal allows to separate the contribution due to the needle movement and the needle closing time can be determined with a higher level of accuracy. In a real application, a reference voltage signal can be either generated and implemented during the calibration phase of the engine or periodically generated by the ECU in order to take also into account the aging effects of the injector.

Then, the signal obtained by the difference between the coil voltage signal and the reference voltage signal can be processed to calculate the time of its maximum value (or minimum, depending on the sign of the difference). This time corresponds to the injector needle closing time.

Being able to measure the time of needle closing means being able to measure the injected fuel mass. In fact, it can be demonstrated that for solenoid injectors, when there is no variation in opening delay, fuel mass amount is linearly correlated with needle closing time. An experimental correlation between needle closing time and the corresponding amount of fuel injected is depicted in Fig. 6.4 for two commercial GDI injectors.

The graph clearly shows that a linear correlation exists between injection duration (i.e., the time elapsing from the starting of the electrical command to the complete needle closing) and the corresponding mass of fuel injected. This correlation does not depend on the particular injector (of the same family) used, and it is also valid at ballistic.

On the basis of this correlation, the control system, once that injection duration has been detected from the voltage signal, can calculate the actual amount fuel amount injected. This fuel mass value is subsequently compared with a target value previously defined, and as a result of this comparison, a correction value for the coil energizing time is defined.

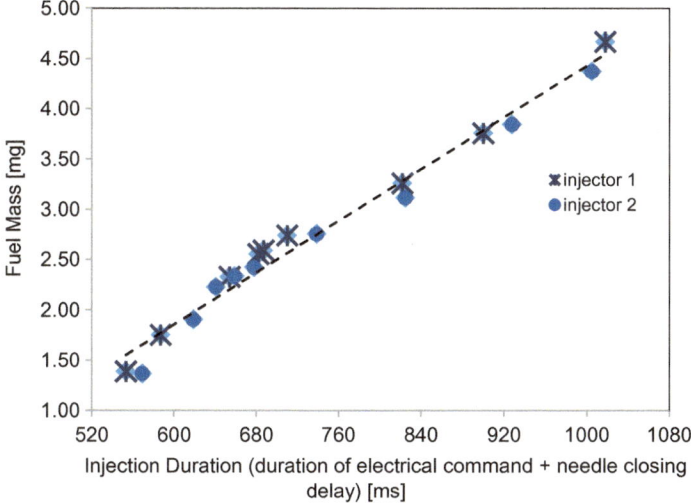

Fig. 6.4 Experimental correlation between injection time duration and delivered fuel mass for two commercial GDI injectors

References

1. C.H. Schwarz, E. Schunemann, B. Durst, J. Fischer, A. Witt, Potential of the spray-guided BMW DI combustion system. SAE Technical Paper 2006-01-1265 (2006). doi:10.4271/2006-01-1265
2. J. King, L. Schmidt, J. Stokes, J. Seabrook, F. Nor, S. Sahadan, *Multiple injection and boosting benefits for improved fuel consumption on a Spray Guided Direct Injection gasoline engine.* Proceedings of the FISITA 2012 World Automotive Congress (2012)
3. M. Parotto, S. Sgatti, F. Sensi, Advanced GDI injector control with extended dynamic range. SAE Technical Paper 2013-01-0258 (2013). doi:10.4271/2013-01-0258
4. M. Parotto, S. Sgatti, F. Sensi, Advanced GDI injector control with extended dynamic range. SAE Technical Paper 2013-01-0258 (2013). doi:10.4271/2013-01-0258

Index

© The Author(s) 2018
F. Taglialatela Scafati et al., *Nonlinear Systems and Circuits in Internal Combustion Engines*, SpringerBriefs in Nonlinear Circuits,
https://doi.org/10.1007/978-3-319-67140-6